PLANT PATHOGENS AND THEIR CONTROL IN HORTICULTURE

Other Related Books

Ayurvedic Medicinal Plants of India, Vol. 1-2 (Set)	Bhutiya, R.K.
Bacterial and Viral Disease and their Management in Horticultural Crops	Reddy, P.P.
Bamboo Flowering and Rodent Outbreaks	Sridhara, S.
Biofuels : Potential and Challenges	Pandey, A.K.
Biofumigation and Solarization for Management of Soil-Borne Plant Pathogens	Reddy, P.P.
Biointensive Integrated Pest Management in Horticultural Ecosystems	Reddy, P.P.
Chickpea (Gram) Status and Cultivation Technology	Shekhawat, S.S.
Controversial Herbal Drugs of Ayurveda	Saroya, A.S.
Diseases of Horticultural Crops: Nematode Problems and their Management	Reddy, P.P.
Flavouring and Fragrant Resources of India	Sood, S.K.
Flowers and Vegetables of India	Sud, R.K.
Folk Herbal Medicine and Drug Discovery	Katewa, S.S.
Fungal Diseases and their Management in Horticultural Crops	Reddy, P.P.
Hort-Agro Project Reports	Sud, R.K.
Horticulture and Livelihood Security	Nath, P.
Insect, Mite and Vertebrate Pests and their Management in Horticultural Crops	Reddy, P.P.
Integrated Nematode Management in Horticultural Crops	Reddy, P.P.
Ornamental Plants for Gardening	Chopra, V.L.
Plant Protection in Horticulture Vol. 1-3 (Set)	Reddy, P.P.
Productivity Enhancing Technologies for Horticultural Crops	Reddy, P.P.
Recent Trends in Disease Management of Fruits and Seeds	Prasad, M.M.
Salinity Tolerance in Plants: Methods, Mechanisms and Management	Garg, B.K.
Science of Tea Technology	Ahuja, P.S.
Silvipasture in India: Present Perspectives and Challenges Ahead	Palsaniya, D.R.
Tropical Forest Ecosystems Structure and Function	Singh, V.P.
Tropical Planting and Gardening 6th Edition	Barlow, H.S.
Useful Herbs of Planet Earth	Daniel, M.
Vegetable Gums and Resins	Howes, F.N.
Weed Seed Biology	Vanangamudi, K.

PLANT PATHOGENS AND THEIR CONTROL IN HORTICULTURE

G.R. DIXON
B.Sc. (Horticulture), Ph.D.
Head of Horticultural Division,
School of Agriculture, Aberdeen, UK

United Book Prints

First Reprint in India
United Book Prints
4806/24, Ansari Road, Daryaganj
New Delhi - 110 002
Tel.: +91-11-41511055
E-mail: unitedbookprints@gmail.com

Print, 2021

ISBN: 978-93-83692-00-2
Reprint : 2014

Reprint in India

CONTENTS

PREFACE

This book describes the application of plant pathology to horticulture. In such a context, control of plant pathogens is viewed as an integral part of crop production. From this point of view, the primary requirement is to produce fruitful, high-yielding crops in the most cost-effective manner rather than simply to control plant pathogens. Cost-effectiveness, however, needs careful definition so that long-lasting effects are achieved with minimal and acceptable disturbance to our environment. This requires knowledge of pathogen biology and of the pros and cons of specific control methods. For this reason the book is divided into two main sections. The first four chapters describe those groups of micro-organism that cause disease; these are accompanied by examples of particular fungal, viral and bacterial pathogens. Chapters 5–10 describe control by means of host resistance, chemicals and spray application, plant health regulation, husbandry and biological methods and control in the post-harvest phase. The examples of pathogens given in chapters 2–4 are expanded upon in chapters 5-10. To achieve this there is cross-referencing between chapters and particular terms are highlighted within the text in order to provide readers with readily available definitions. Inevitably in a short book, simplifications are essential; these may offend the pedant but provide the student reader with basic information. On this he or she can build and delve more deeply into facets of plant pathology. The latter purpose is served by extensive lists of references and further reading. Examples of plant pathogens have been limited but reflect the importance of particular groups such that there are several thousand fungi, several hundred viruses and only one or two hundred bacteria pathogenic to crop plants.

Books in the Science in Horticulture Series are intended for a wide range of students from diploma to undergraduate status. While this book is intended for students of horticulture it will also be valuable for those taking plant pathology and microbiology modules in agriculture, botany and forestry courses. The wider audience of growers, advisers, researchers in related fields and knowledgeable amateurs is provided with an up-to-date account of plant disease control. To help readers, some topics are broken down into short subsections in order to convey the gist of a subject area.

Valuable comments and criticism on all or part of this text have been made by: Professor L. Broadbent (Series Editor), Dr S. A. Hill, Dr P.

Howell, Mr D. R. Humphreys-Jones, Mr A. D. McKelvie, Dr A. Paton, Mr E. B. Scragg, Dr M. Shaw, Dr S. Wale and Dr A. J. Whalley. All errors and omissions are my own fault. Several essential sources of reference have been provided by Dr D. M. Spencer. I record my thanks to those who have so freely provided illustrations; specific acknowledgements are given at appropriate points in the text. I am especially grateful to Mrs M. Sinclair who has cheerfully and efficiently typed my longhand into intelligible script; Mrs A Park and Mrs M. Forbes have provided great assistance in tracing reference material. Particular thanks are due to my wife, Kathy, and to Lucy and Richard for their forbearance and understanding during the preparation of this book. Staff at the Macmillan Press, especially Mr M. J. Stewart, have been a constant source of help and encouragement.

Aberdeen, 1983 G. R. D.

1 PATHOGENS AND DISEASES

1.1 CROP LOSSES CAUSED BY PATHOGENS

The term *pathogen* indicates an ability on the part of a micro-organism to cause disease which results from an intimate relationship with the host plant. In this process the pathogen derives energy for growth and reproduction without any benefit to the host. *Disease* is, in consequence, a change in the normal functioning of plants and results in poorer growth and lower yields. It is of primary consideration for those concerned with the control of plant pathogens to establish the amount and timing of crop loss associated with particular levels of pathogen infection. This permits an evaluation of the methods for pathogen control discussed in chapters 5–10 from which a balanced judgement can be achieved of their costs and benefits. This type of analysis has been done with considerable precision for some temperate cereal crops. For instance, Large and Doling (1962) established that the square root of the percentage leaf area infected with *Erysiphe graminis* (powdery mildew) is related to grain yield lost in tonnes per hectare. Horticultural crops, however, have a combination of yield characters which often makes such direct estimations difficult. The value of these crops frequently rests on quality characters such as visual appeal as with flowers, fruit, pot plants, trees, shrubs and vegetables. Alternatively, value may lie in an estimation of future productivity as with transplanting material sold for growing-on or forcing. Land areas devoted to individual crops are relatively small and often widely dispered around the country. This tends to confound the agricultural yardstick of area statistics for assessing the importance of crops. The cash value of individual plants is extremely high. For example, with glasshouse tomatoes a 1 per cent reduction in yield with plants grown at 25 000 plants ha^{-1} and each worth £6 could represent a loss of £1500 ha^{-1} to the producer. Values for container-grown hardy nursery stock are far higher and concomitantly losses could exceed those of tomatoes. For 3-year-old *Chamaecyparis lawsoniana* 'Ellwoodii' valued at £2.50 each and grown on sand beds at 16 plants m^{-2} a loss of 10 per cent due to disease represents a loss of £4 m^{-2}. For some crops such *pro rata* calculations may not be accurate. A 10 per cent infection by *Erysiphe cruciferarum* (powdery mildew) of the bud area of Brussels sprouts valued at £7500 ha^{-1} could mean that the entire crop is rejected for processing, meaning total crop failure, or that the produce is downgraded in the open market.

Slight blemishes can diminish the value of a crop by affecting its visual quality. These may even be the result of resistance reactions, for example, ghost spotting of tomatoes caused by single spores of *Botrytis cinerea* (grey mould—section 2.6.4) germinating on the fruit surface but being unable to penetrate and establish a colony. This has no effect on the eating quality of the fruit but makes it unacceptable to a supermarket buyer because of lost visual quality. Blemished material has to be removed manually from fruit and vegetables used for processing. Processing companies freely admit that with strawberries, raspberries and green beans they would be prepared to pay higher prices for guaranteed total freedom from blemishes caused by *B. cinerea*.

The intensity and continuity of horticultural crop production may be severely limited by both soil-borne and aerially dispersed pathogens. These may affect the intensity of time-scale whereby several crops are grown in one season on the same land. A build-up of *Sclerotinia sclerotiorum* in soil can prevent land being cropped with carrots, dwarf beans, lettuces and parsnips. There can be very precise requirements for crops to mature on specific dates. Pathogens can wreck this type of scheduling. For example, *Peronospora viciae* (downy mildew) causes the pea crop to ripen unevenly and makes harvesting difficult. This usually results in the whole crop being rejected for freezing because it gets out of synchrony with previously agreed production schedules and is fit only for use as seed or dried peas.

There are various forms that crop loss may take in horticultural crops.

Direct yield loss

This is a measurable reduction in tonnes per hectare or crates per hectare of fruit or field vegetables—or, under protection, number of blooms per square metre. Elegant studies of yield depression in apples affected by chat-fruit disease were made by Posnette and Cropley (1965) and are illustrated in figure 1.1. Over a seven-year period there was approximately an 8 per cent reduction in mean fruit weight irrespective of whether infection originated from the rootstock or scion. Since apple grades are closely related to fruit weight this would produce largely second-rate fruit from infected trees which, in the present highly competitive top-fruit market, would be virtually unsaleable.

Rejection

Either the whole crop or perhaps only part of it may be rejected, and this can be influenced by a range of market factors. Buyers for supermarket chains may reject lettuces with only single lesions of *Bremia lactucae* (downy mildew—section 2.6.3) in glut periods, but in times of scarcity 5 per cent leaf area infection could be accepted. Factors that influence

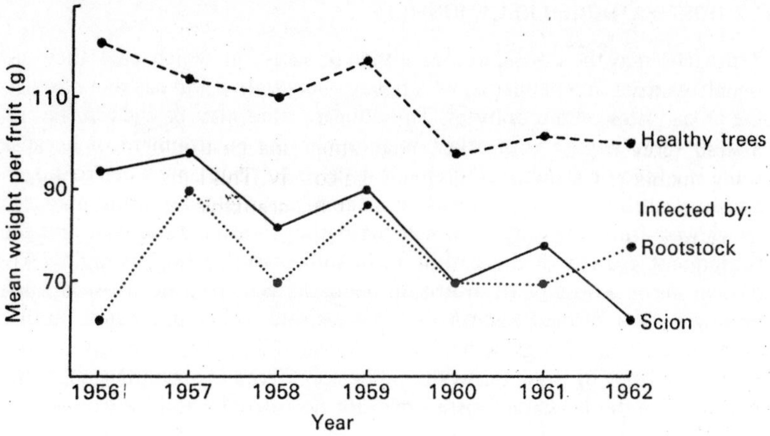

Figure 1.1 Effect of chat-fruit disease on fruit size of Lord Lambourne apple trees on M.VII rootstocks (after Posnette and Cropley, 1965). Reproduced by permission of East Malling Research Station and Association of Applied Biologists

this decision will be: time of year, weather (is there a demand for salads?) and the availability of alternatives such as celery, chicory or cress.

Change of grade
Otherwise sound produce may be downgraded by the Market Inspectorate because it has been blemished by a pathogen. Disease criteria form part of the European Economic Community grading schemes for crops such as apples, pears, carrots, cucumbers, sprouts and tomatoes.

Shifts in harvest period
Abortion of early season trusses of tomato fruit caused by infection with tomato mosaic virus leads to later yields when prices are falling because larger volumes of fruit are being marketed.

Alteration of crop rotation
Entry of *Plasmodiophora brassicae* (clubroot—section 2.6.1) on to land used for intensive brassica production means that areas of a holding have to be isolated to prevent further spread of the pathogen. This complicates the planning of rotations and reduces the value of a holding for brassica vegetable production.

1.2 HOST-PATHOGEN RELATIONSHIPS

Pathogens may be considered as either *obligate*, in which case they are unable to exist independently of a host, or *facultative* and capable of existing as parasites or saprophytes. The obligate state may be the result of an inability to compete with other organisms in the environment or because there is a specific growth requirement which is fulfilled only when pathogen and host are in intimate association. A host–parasite relationship may also be viewed from the way in which the pathogen affects the host and the subsequent growth of the pathogen. In some cases pathogens rapidly kill part or all of a host plant, nutrition being derived from dead tissue; such pathogens are termed *necrotrophs*. Conversely, other pathogens require living host tissue for growth and development and these are termed *biotrophs*. Neither of these terms should be considered as absolute since some pathogens exhibit characteristics of both necrotrophs and biotrophs. The ultimate biotrophs are viruses (chapter 3) which are incapable of metabolism outside a host cell. Some fungi such as the powdery and downy mildews and the rusts (chapter 2) appear to have strictly biotrophic requirements whereas those bacteria that cause soft rot diseases (chapter 4) are extreme necrotrophs. Between these states are organisms such as potato blight fungus, *Phytophthora infestans* (section 2.6.2). This pathogen has a high degree of host specificity (restricted to potato, tomato, green pepper, aubergine and some wild *Solanum* spp.) and grows in intimate association with the host (biotrophic characteristics) but causes rapid death of host tissues (necrotrophic character). The main features of biotrophs and necrotrophs are shown in table 1.1.

1.3 FORMS OF DISEASE EXPRESSION

The effects of a pathogen on a host are often first seen as *chlorosis*, loss of the normal green colouration with affected organs turning yellow-green, yellow or white, followed by *necrosis* when there is death of part of a host or of a tissue. A range of colloquial terms is applied to plant diseases which briefly describe some but not all of the associated host symptoms (the totality of symptoms being known as a *syndrome*). In some cases these terms may also be used to group diseases that are characterised by similar symptoms and in this sense, as well as for means of general description, these terms have value. It should be borne in mind, however, that diseases within such a group may be caused by different types of micro-organism and in consequence require dissimilar control measures. The following are examples of some of the more generally applicable terms; a more comprehensive list is given in Anon (1973). They may be divided into terms that are indicative solely of a host reaction and those that also include the pathogen.

Table 1.1

Major characteristics of necrotrophic and biotrophic pathogens (Dickinson and Lucas, 1982)

*Necrotrophs**	*Biotrophs**
Biochemical and morphological features	
Host cells rapidly killed	Host cells not rapidly killed
Toxins and cytolytic enzymes produced	Few or no toxins or cytolytic enzymes produced
No specialised parasitic structures produced	Special parasitic structures such as haustoria typically formed
Host penetration via wounds or natural openings	Host penetration direct or via natural openings
Ecological features	
Wide host range	Narrow host range
Able to grow saprophytically away from the host	Unable to grow away from the host
Attack juvenile, debilitated or senescing tissues	Attack healthy hosts at all stages of development

*The terms necrotroph and biotroph were proposed by Gäumann (1950)

1.3.1 Host expression

Anthracnose

This literally means 'like coal or charcoal'. These diseases are characterised by a development of black lesions, usually sunken, such as those found on pods of *Phaseolus* bean infected by the pathogen *Colletotrichum lindemuthianum*.

Blight

This term is commonly applied to diseases where there is devastatingly rapid destruction of host tissue and is probably best reserved for diseases with such catastrophic effect rather than those involving leaf spotting. One of the best examples is potato blight caused by *Phytophthora infestans* (section 2.6.2). Others found on horticultural crops include: raspberry cane blight caused by *Leptosphaeria coniothyrium*, bacterial blights caused by *Erwinia amylovora* (fire-blight), *Pseudomonas mors-prunorum* and *P. syringae* (canker and shoot blight) and chrysanthemum ray and petal blights caused by *Didymella chrysanthemi* and *Itersonilia perplexans*.

Canker

Initially the host bark becomes discoloured, darkened and sunken. The cortex and phloem dry out, separate from the wood tissue and eventually crack, breaking away and exposing the xylem; successive folds of callus tissue form around the area of necrosis. This type of disease is seen most commonly on ornamental and orchard shrubs and trees, for example, bacterial canker caused by *Pseudomonas mors-prunorum* and *P. syringae*.

Damping-off

This symptom is usually associated with the collapse and subsequent death of seedlings following the formation of stem lesions at soil level. It is also associated with water-soaking, browning or shrivelling of stem tissues at soil level. The plants usually fall over and die. 'Damp' refers to conditions of high humidity within the plant stand which favour infection by and spread of a pathogen. With this term may also be included pre-emergence death of seedlings as indicated by poor erratic emergence. A pathogen that causes damping-off of a wide range of hosts is *Thanatephorus cucumeris* while a more restricted host range is affected by *Aphanomyces cochlioides* and *Pleospora björlingii*.

Gall

Localised swellings or outgrowth of host tissue may be termed galls and result from *hypertrophy* (excessive enlargement of individual cells) and *hyperplasia* (excessive multiplication of cells). Galls may result from the indirect effects of a pathogen on host hormone metabolism. Examples include clubroot caused by *Plasmodiophora brassicae* (section 2.6.1) and crown gall caused by *Agrobacterium tumefaciens*.

Mosaic

Frequently mosaic symptoms result from infection by a virus pathogen; they generally affect leaves and are seen as numerous small areas of discolouration against a background of a different tint with defined boundaries delineated by the veins. Examples are diseases caused by cauliflower mosaic virus (section 3.5.1), lettuce mosaic virus (section 3.5.2), apple mosaic virus (section 3.5.6), common bean mosaic virus, cucumber mosaic virus and tomato mosaic virus.

Mottle

A pattern of small numerous zones of discolouration is termed a mottle. Often the zones are chlorotic with irregular shapes and lacking distinct boundaries against a background of a different tint; their distribution is unrelated to the vein network. An example is the effects of carnation mottle virus (section 3.5.5).

Scab

Abnormal thickening of the host surface layers possibly associated with cork-tissue formation and perhaps localised excessive host cell proliferation leads to the development of a scab. This term is applied also to those diseases that involve a development of discrete superficial lesions with localised surface roughening or pitting. Examples are apple scab caused by *Venturia inaequalis* (section 2.6.5) and common scab of potato caused by *Streptomyces scabies*.

Spot

Spots are typified by the development of separate, isolated lesions roughly circular in shape, differing in tint from surrounding tissue and often chlorotic or necrotic. This term may be qualified as for example in *ringspot* where there are distinct concentric zones of colour composing the lesion. Examples of spot diseases are ringspot of raspberry caused by a virus pathogen (section 3.5.4), dark leaf and pod spot of brassicas caused by *Alternaria brassicae* and *A. brassicicola*, celery leaf spot caused by *Septoria apiicola* and rose black spot caused by *Diplocarpon rosae*.

Rot

This is a term frequently applied to plant diseases; it indicates tissue disintegration and may be qualified as brown, collar, dry, foot, neck, root, soft, wet and white. Examples are pea and bean root rots caused by *Aphanomyces euteiches*, *Fusarium solani* f. sp. *phaseoli* and *F. solani* f. sp. *fabae*, root rots of a range of hosts caused by *Pythium* spp., *Helicobasi-*

dium purpureum violet root rot, storage rot caused by *Mycocentrospora acerina*, root rot of beet caused by *Pleospora björlingii*, onion rot caused by *Botrytis* spp., onion white rot caused by *Sclerotium cepivorum*, corky root rot of tomato caused by *Pyrenochaeta lycopersici*, *Narcissus* basal rot caused by *Fusarium oxysporum* f. sp. *narcissi*, root and foot rot of a range of hosts caused by *Phytophthora cinnamomi* and butt and heart rot of conifers caused by *Heterobasidion annosum*.

Wilt

Symptoms associated with wilt diseases are a loss of turgidity and foliar collapse resulting either from pathogenic damage to the host root system or invasion of the vascular system. Particular symptoms of specific wilt diseases vary with the causal organism but often an initial sign is downward bending of the lower leaf petioles (*epinasty*) plus slight vein clearing and yellowing of these leaves. Affected leaves gradually become more chlorotic and die, while symptoms are expressed progressively up the plant. Initially wilting may be reversible, with the host recovering at night, but eventually loss of turgor becomes permanent with resultant shrivelling and death. Where diseases are caused by pathogens localised in the vascular system, tyloses and gum deposits may be found in the xylem vessels accompanied by collapse of some vessels and disintegration of adjoining parenchyma tissue. Examples are wilt diseases caused by the fungi *Verticillium albo-atrum* and *V. dahliae* (section 2.6.9) and *Fusarium oxysporum* (on a wide range of crops), *Ceratocystis ulmi* (Dutch elm disease), by bacteria such as *Corynebacterium michiganense* (tomato wilt) and viruses (for example, tomato spotted wilt).

1.3.2 Pathogen expression

There are some terms that have several shades of meaning in that they describe both manifestations of pathogen growth on a host and symptoms. More correctly they are used to describe groups of biologically related pathogens causing similar symptoms. Examples of these terms are as follows.

Mildew

Other terms, such as 'downy' or 'powdery', are used to qualify mildew. Infection by a downy mildew is often first seen as discrete fungal sporulations usually on the undersurfaces of leaves or other organs developing as a white to grey 'bloom' in humid conditions. The causal pathogen grows within the host tissues forming fruiting structures through the stomata and these give rise to the 'downy' effect. Examples of pathogens that produce this effect are *Peronospora parasitica* (brassica downy mildew), *Peronospora viciae* (pea downy mildew) and *Bremia lactucae* (lettuce downy mildew—section 2.6.3). In contradistinction, powdery mildews grow on the host surface, usually the upper surface, and fungal growth together with copious formation of conidia gives a 'powdery' effect. Examples of pathogens that produce this effect are *Sphaerotheca fuliginea* (cucumber powdery mildew), *S. macularis* (powdery mildew of strawberry and raspberry), *S. mors-uvae* (American gooseberry mildew), *S. pannosa* (rose powdery mildew—section 2.6.6), *Podosphaera leucotricha* (apple powdery mildew) and *Microsphaera alphitoides* (oak powdery mildew).

Rust

This term is derived from the presence of orange-brown spore masses on the host surface which originate from growth of some rust fungi within the host tissue. Examples of rust pathogens are *Cronartium ribicola* (blister rust of *Pinus* and *Ribes* spp.—section 2.6.8) and *Uromyces dianthi* (carnation rust).

REFERENCES

ANON (1973). A guide to the use of terms in plant pathology. *Phytopathological Papers*, No. 17. Commonwealth Mycological Institute, Kew

DICKINSON, C. H. and LUCAS, J. A. (1982). *Plant Pathology and Plant Pathogens*, Blackwell, Oxford

GÄUMANN, E. (1950). *Principles of Plant Infection* (translated by W. B. Brierley), Hafner Co., New York

LARGE, E. C. and DOLING, D. A. (1962). The measurement of cereal mildew and its effect on yield. *Plant Pathology*, 11, 47–57

POSNETTE, A. F. and CROPLEY, R. (1965). Field experiments with chat fruit virus disease of apple. *Annals Applied Biology*, 55, 439–45

FURTHER READING

ANDERTON, P. and BIGG, P. H. (1972). *Changing to the metric system, conversion factors, symbols and definitions*, Her Majesty's Stationery Office, London

COOKE, R. C. (1977). *Fungi, Man and His Environment*, Longman, London

HENDERSON, I. F. and HENDERSON, W. D. (1963). *A Dictionary of Biological Terms* (8th edn, ed. J. H. Kenneth), Oliver and Boyd, Edinburgh

PRIMROSE, S. B. (1975). *Introduction to Modern Virology*, Blackwell, Oxford

SMITH, K. M. (1977). *Plant Viruses*, Chapman and Hall, London

WHEELER, B. E. J. (1969). *An Introduction to Plant Diseases*, Wiley, London

WILKINSON, J. F. (1975). *Introduction to Microbiology*, Blackwell, Oxford

WOOD, R. K. S. (1975). *Disease in Higher Plants*, Oxford University Press, London

2 FUNGI

2.1 THE NATURE OF FUNGI

Fungi are living organisms that are devoid of chlorophyll, obtaining energy either as parasites or saprophytes. Simple organic compounds such as amino acids and monosaccharides are absorbed directly. More complex materials are degraded first by fungal-produced extracellular enzymes and then absorbed. Parasitic fungi must be capable of overcoming host resistance mechanisms in order to obtain supplies of energy. Many fungi are filamentous and multicellular although some are single cellular. Fungal nuclei are small rounded bodies, usually 2–3 μm in diameter (with some species they are up to 30 μm in diameter) and bounded by a double layer of membranes with conspicuous pores. Characteristically, the nuclear membranes persist throughout nuclear division and do not disappear at prophase as in other organisms. The nucleolus is also retained during division in some fungi. Typically, a vegetative fungal *thallus* (a tissue of cells that has little or no differentiation) is formed of filaments which branch in all directions. Each filament is known as a *hypha* and an aggregation of hyphae is known as a *mycelium*. Hyphae are thin transparent tubes containing protoplasm and bounded by a rigid cell wall. The walls of fungal hyphae are composed of many layers (*multilaminate*) each consisting of variously orientated fibrils. Characteristically, fungal cell walls contain chitin. This is a simple straight chain molecule resembling cellulose except that the basic unit of the chain is N-acetyl glucosamine, instead of glucose; this material is generally absent from higher plants. Other compounds such as proteins and lipids, found throughout the plant kingdom, are also present in fungal walls. The protoplasm has a general structure similar to that of other eucaryotes, containing organelles and inclusions such as mitochondria, vacuoles, vesicles, endoplasmic reticula, ribosomes, microbodies, microtubules, crystals and glycogen. Lomasomes are more common in fungi than in most other organisms whereas Golgi bodies (or dictyosomes) are apparently less frequent. Cross walls (*septa*) are found at irregular intervals in the hyphae of Ascomycotina, Basidiomycotina and Deuteromycotina (section 2.5). In the morphologically simpler types such as Oomycetes and Zygomycotina, cross walls are generally absent except at the bases of reproductive hyphae or in senescent ones. Hyphae that lack septa are described as *coenocytic*. The septa are perforated and consequently protoplasm is continuous throughout each hypha. Extension occurs only at the extreme tip of each hypha. *Apical growth* of this type is a diagnostic feature of the fungi and distinguishes them from most other organisms. The apical cell or compartment of a hypha contains several nuclei together with cytoplasmic organelles. At the extreme tip these are replaced by an accumulation of membrane-bound vesicles which probably have an essential role in growth. Compartments further back along the hypha may each contain one or more nuclei. Also present are membrane-bound vacuoles which are small in young cells but enlarge and appear to coalesce in older cells, that is, those further back from the apex. Eventually they restrict the rest of the cytoplasm to a thin peripheral zone. In the oldest parts of a hypha, cytoplasm may be withdrawn altogether from some compartments and accumulate in others (Beckett *et al.*, 1974). At times in the life history of some fungi the hyphae become woven into compacted tissues. If the hyphae are loosely interwoven and remain easily distinguishable this tissue is called *prosenchyma*; where the hyphae are more closely packed and cannot be differentiated the tissue is known as *pseudoparenchyma* and is analogous to the parenchyma of vascular plants. Both tissue types form vegetative and reproductive structures in the fungi.

Fungal growth occurs between 0–35°C with a general optimum of 20–30°C. Fungi grow best in acidic conditions with a general optimum pH of 5–6.5 (compare with bacteria, see chapter 4). Light is not required for growth in most fungi but in some species certain wavelengths encourage sporulation. Most fungi are aerobic in their mode of respiration, some are capable of anaerobic respiration but none are obligately anaerobic. Owing to the thin-walled nature of most fungal structures there is a requirement for high levels of relative humidity (RH) or the presence of free water without which desiccation quickly takes place. This constraint is circumvented by the production of thick-walled resting and reproductive structures which permit dormancy during dry periods (section 2.3). A few fungi such as some members of the Erysiphales are favoured during parts of their life cycle by conditions of low RH. The growth pattern of fungi on a nutrient substrate is typical of that of other micro-organisms and follows the curve shown in figure 2.1.

2.2 FUNGAL REPRODUCTION

Fungi produce several types of *spore*. These are specialised reproductive cells that do not contain an embryo. They originate by either sexual or asexual processes and are formed in prolific numbers. Sexual spores tend either to be capable of resisting adverse environmental conditions, that is, they have a survival function, or to develop after a period of adverse conditions and are essential for the establishment of primary infections

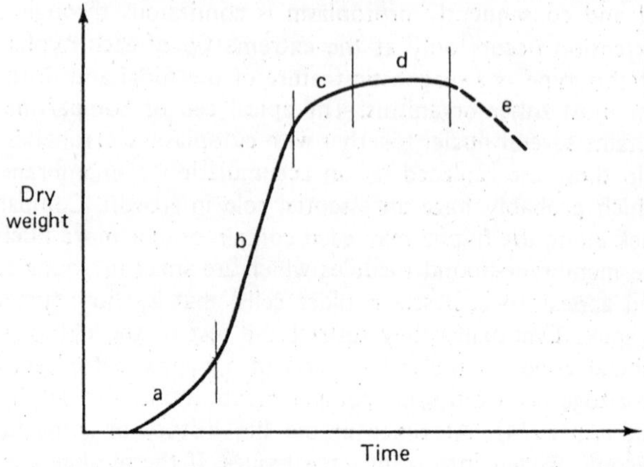

Figure 2.1 Growth pattern of a fungus on a nutrient substrate. (a) Lag phase: no cell multiplication, adaptation to growth in a new environment. (b) Exponential (log) phase: cells growing at a constant rate; the generation time (period between divisions) is constant, leading to an exponential increase in the mass and number of cells. (c) Deceleration phase: gradually the growth slows as nutrients and space become limiting factors. (d) Plateau phase: cell number becomes constant, toxic products may accumulate. (e) Death phase: cells begin to die and there is a difference between total number of cells and number of viable cells; cells may digest themselves (autolysis)

on new host growth or in new seasons. Asexual spores are distributed rapidly at times that coincide with host availability. This leads to rapid expansion of pathogen populations and quick colonisation of crops during periods of favourable weather and the development of *epidemics* (a widespread temporary increase in the incidence of an infectious disease). Asexual spores are likely to be formed from a mycelium several times in one season; this increases the rate of spread of a pathogen, whereas sexual spores may occur only once during a single season.

2.2.1 Asexual reproduction

Asexual reproduction is broadly defined as any means by which fragmentation of an organism occurs whereby each fragment is capable of growing to become a new individual. Four means of asexual reproduction may be identified: production of spores each of which may germinate to produce a new individual, fragmentation of the thallus, budding of somatic cells where each bud forms a new individual and fission of adult cells into daughter cells. The first three are found in the fungi; the fourth is more common in bacteria. Spore production is the most common form in the fungi, leading to a production of sporangiospores and conidia.

Sporangiospores

Spore sacs (*sporangia*) are borne on specialised hyphae (*sporangiophores*) the contents of which cleave forming sporangiospores. These may be either motile and known as *zoospores* actively dispersing in moisture films, or non-motile known as *aplanospores* and dispersed passively. In conditions of low RH, sporangia may germinate via production of a germ tube; some sporangia, for example those of *Bremia lactucae* (lettuce downy mildew—section 2.6.3), preponderantly germinate in this way irrespective of environmental conditions.

Conidia

These form at the tips or on the sides of hyphae (*conidiophores*) in various ways. In some cases the conidiophores are hyphal branches with a specialised asexual function. The conidia usually develop as discrete chains of cells; there is no cytoplasmic cleavage in their formation and dispersal is passive. Germination takes place through the formation of a germ tube. Conidia may be classified according to their modes of formation (Kendrick, 1971).

Where *fragmentation* takes place the hyphae break up into their component cells which then behave as spores (*arthospores*). In some cases there may be thickening of terminal or intercalary hyphal segments which become pigmented, accumulate food reserves and are termed *chlamydospores*. Usually there is no mechanism for separation or dispersal of chlamydospores except the disintegration of surrounding hyphae. They are an important survival mechanism for soil-borne fungi.

Budding is the production of a small outgrowth from a parent cell. The nucleus of the parent cell divides in unison with bud formation and one daughter nucleus migrates into the bud. The latter increases in size while still attached to the parent but eventually breaks away forming a new individual. Asexual reproduction by budding is common in yeasts but is found in other fungi. *Fission* is the simple splitting of one cell into two daughter cells by constriction and formation of a cell wall; it is found in some yeast fungi.

Fungi can often reproduce solely by asexual means, thereby repeating the stage in the life cycle from which asexual spores are produced. In the Deuteromycotina (*Fungi Imperfecti*) reproduction is solely by asexual means. The structure, shape, colour and mode of formation of asexual spores are important diagnostic features for fungal identification.

2.2.2 Sexual reproduction

Sexual reproduction in fungi, as with other organisms, involves the union of two compatible nuclei. Six phases may be identified in this process.

1. *Formation of sex organs.*
2. *Plasmogamy (protoplast fusion)*, which unites two haploid nuclei each possessing x chromosomes together in a single cell. It should be noted that many fungi are haploid for the greater part of their life cycle.
3. *Karyogamy (nuclear fusion)* when the two haploid (x) nuclei fuse into a single diploid $2x$ nucleus (*zygote*).
4. *Meiosis*, the process that restores the haploid x condition forming four nuclei which result from chromosome replication and division.
5. *Development of fruiting bodies* to disperse the sexual spores.
6. *Development of sexual spores.*

The sexual organs of fungi are known as *gametangia*, male organs being *antheridia* and female being *oogonia*. On the basis of how these organs are carried, three types of fungi may be differentiated.

1. *Hermaphroditic*—where each thallus carries male and female organs which may or may not be compatible with each other.
2. *Dioecious*—where the male and female organs are borne on different thalli.
3. *Sexually undifferentiated*—where sexual structures are morphologically indistinguishable as male or female.

Where hyphae within a mycelium cross, they may fuse (*anastomosis*) and there is then localised cell-wall breakdown with subsequent nuclear and cytoplasmic exchange. In the Basidiomycotina (section 2.5) anastomosis permits nuclei of different types to migrate into the same hypha. These nuclei do not fuse immediately, consequently a *dikaryon* is formed which, since the nuclei are genetically dissimilar, is also a *heterokaryon*. Fruiting bodies composed of dikaryotic hyphae then develop. Similar events occur in some Ascomycotina (section 2.5) but the dikaryotic phase is restricted to fertile hyphae that are found within the fruiting body but not in the mycelium as a whole. Some fungi possess an ability for *parasexual reproduction* whereby plasmogamy, karyogamy and meiosis take place but not in specialised sexual organs or at any specific point in the life cycle (section 5.4). Both heterokaryosis and parasexuality promote genetic variability within the fungi and have an important effect upon the usefulness of host resistance as a means of pathogen control (section 5.4).

The following five spore forms result from sexual reproduction in the fungi.

Oospores

These are resting spores having a survival function, possessing a thick wall and deposits of lipid food reserves. They have limited dispersal value. Oospores are particularly characteristic of the Oomycetes (Mastigomycotina) and develop from gametes of markedly different sizes with the oogonia being far larger than the antheridia.

Zygospores

These are similar to oospores, are characteristic of the Zygomycotina and result from plasmogamy and karyogamy between gametangia of equal size.

Ascospores

These are characteristic spores of Ascomycotina resulting from nuclear fusion followed by meiosis. Normally four haploid daughter nuclei divide forming eight nuclei around each of which cytoplasm aggregates to produce ascospores in a cylindrical sac (*ascus*). They are ejected from the ascus into air currents, have thin walls and limited food reserves and are fitted for dispersal but not perennation.

Basidiospores

These are characteristic spores of the Basidiomycotina. They are typically unicellular, although there are exceptions, haploid and are borne externally on a club-shaped hypha (*basidium*). Basidiospores result from karyogamy and meiosis, followed directly by mitosis in many species. They are thin-walled, dispersed by air currents and have a limited capacity for survival.

Resting Sporangia

These are thin-walled structures found embedded within host tissue and are characteristic of some members of the Chytridiomycetes.

2.3 DISPERSAL AND PERENNATION

Sporulation, resulting from either sexual or asexual processes, is the key to movement and survival in fungi and hence to the establishment of new infections and to perennation between hosts. As can be seen from section 2.2, the fungi produce a wide diversity of spores each adapted for dispersal and/or perennation. This contrasts markedly with the bacteria where, with a few exceptions, spore form is extremely uniform.

Colonisation of a host must eventually lead to spread or dispersal as space and nutrients limit further growth and reproduction. An efficient dispersal mechanism to new hosts is a basic requirement for pathogens. The agencies that promote fungal dispersal are wind, water, host plant material (seeds and vegetative parts), soil, man and other animals. Dispersal may take place either very rapidly, particularly with the production of airborne propagules, or over a long period as with soil-borne pathogens. If a vector is involved then dispersal is dependent on it in terms of the speed and distance of dispersal.

Dispersal via the atmosphere has three distinct phases: liberation, dispersion and deposition of the propagules. Some pathogenic fungi have evolved sophisticated methods which propel spores away from the host and through the zone of still air that surrounds the surfaces of soil and plant structures (Ingold, 1971). The thickness of this boundary layer varies with factors such as wind speed and turbulence, and size and shape of the surface. Fungal dispersal units take off cleanly from the parent colony carrying none of the extraneous material often associated with bacterial cells. The process of liberation may be either active or passive. Active liberation is found, for example, in Ascomycotina and Basidiomycotina. The ascocarp may contain explosive asci usually in the form of a squirting mechanism which on bursting can discharge ascospores up to 600 mm, according to species. Initiation of the discharge may be closely dependent on the incidence of rain, as for example, in *Venturia inaequalis* (apple scab—section 2.6.5). In some cases, dew formation may be sufficient to trigger active spore discharge as in *Mycosphaerella pinodes* (pea blight) (Carter, 1963). Many fungal pathogens are liberated passively, for example, uredospores of rusts, conidia of powdery mildews, chlamydospores of smuts and the spores of many conidial fungi such as *Botrytis cinerea* (grey mould—section 2.6.4). In these organisms it is the direct action of wind or leaf-flutter and stem vibration produced by wind or by heavy rain that is chiefly responsible for spore liberation.

It is usual to find a decline in disease development receding from a centre of airborne inoculum production, that is, the degree of infection plotted against distance is a hollow curve. Thus dispersion and deposition are related to increasing distance from a point source of spore liberation. This model is correct over relatively short distances but other factors may complicate the situation over larger distances. It has been established, for example, that uredospores of *Puccinia graminis* (black rust of cereals) can be carried for hundreds of kilometres and lead to outbreaks of rust on wheat in Europe, India, USA and the USSR. It is likely that spores become incorporated in the air masses being moved as 'spore clouds' over long distances.

Deposition of spores by gravity is only important in still air, in turbulent conditions rain is probably a dominant factor. Spore size will influence the rate of fall and it is noted that many aerially dispersed fungi produce large spores by contrast with those of soil-borne organisms such as *Aspergillus* and *Penicillium* spp. Heavy rain will quickly comb spores greater than 5 μm in diameter from the air. Unwettable spores are confined to the surface of a capturing rain drop while wettable ones become distributed throughout it.

Water droplets splashing on to a sporulating mycelium can collect and carry spores. This form of dispersal is essentially a means of short-range transmission but if combined with winds may spread spores quite significant distances. Few fungi are spread by insects but where this does occur a wide range may be involved such as: aphids, bees, beetles, bugs, crickets, moths and wasps. In specific instances a specialised relationship may evolve like that between *Ceratocystis ulmi* (Dutch elm disease) and the elm bark beetle (*Scolytus*).

Only relatively few fungi affect solely subterranean host tissues: those that do form spores that persist for long periods. Disperal is slow therefore, generally with patches of plants affected within an apparently healthy crop. Some of these pathogens are obligate parasites such as *Plasmodiophora brassicae* (clubroot—section 2.6.1) and are unable to propagate when a host is absent whereas others such as *Pythium* and *Thanatephorus* spp. are equally effective saprophytes as parasites which means that spread may take place in soil by hyphal growth. The potential for dispersal (and also dormancy) is increased in some fungi by hyphal aggregation which results in the growth of macroscopic structures such as the following.

Mycelial strands

These are aggregations of parallel, relatively undifferentiated hyphae lacking well-defined apical meristems. They are generally produced from a food base, travel across a nutrient-deficient zone and permit colonisation of a new nutrient source. By cross-growth of strands a network is established. Such structures are found in the Ascomycotina, Basidiomycotina and Deuteromycotina.

Rhizomorphs

In contrast rhizomorphs are highly differentiated with a well-developed apical meristem possessing a central core of large thin-walled cells and an outer layer of small thick-walled cells which are darkly pigmented. Colonisation from host to host by *Armillaria mellea* (honey fungus—section 2.6.7) is achieved by rhizomorphs.

Various environmentally resistant structures that aid fungal perennation have already been described, for example, resting spores, chlamydospores and hyphal aggregations. Some fungi develop *sclerotia* which have a survival function. Sclerotia vary in size from a few mm to several cm and the following four main types have been recognised (Townsend and Willetts, 1954; Willetts and Wong, 1971).

1. *Loose type*. These show little differentiation into zones with the hyphae loosely packed together as found in *Thanatephorus cucumeris* (a cause of damping-off).

2. *Terminal type*. These develop from repeated dichotomous branching of hyphae and are well differentiated into a rind of several layers of rounded, dark and thickened cells, a cortex of thin-walled closely packed cells and a central medulla of loosely arranged filaments as found in *Sclerotinia fuckeliana* (*Botrytis cinerea*, grey mould—section 2.6.4).
3. *Strand type*. These form from hyphae that come to lie parallel to each other and eventually differentiate into a rind and medulla as found in *Sclerotinia (Stromatinia) gladioli* (the cause of corm dry rot in *Gladiolus* and *Crocus*).
4. *Other types*. In some fungi, sclerotia develop from pre-existing mycelia such as that in the ovary of cereals and grasses and which form the ergots of *Claviceps purpurea*.

Sclerotia germinate to form a mycelium, conidia or sexual fruiting bodies such as ascocarps or basidiocarps. Hyphae also aggregate for reproductive functions producing structures in or on which spores develop as in the Ascomycotina, Basidiomycotina and Deuteromycotina. These are discusssed in sections 2.6.4, 2.6.7, 2.6.8 and 2.6.9 respectively.

The means by which perennation takes place are summarised in figure 2.2.

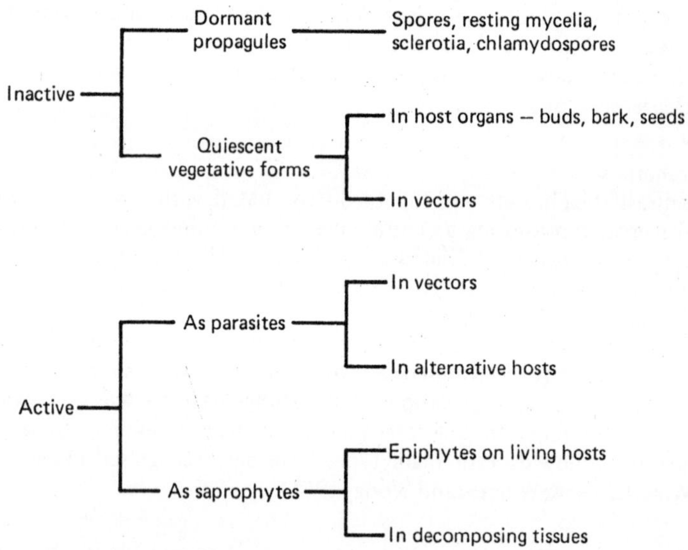

Figure 2.2 Some means of perennation by fungal pathogens (after Dickinson and Lucas, 1977)

2.4 EFFECTS OF FUNGI ON HOST PLANTS

Few fungi penetrate host plants through wounds or natural openings (compare with bacteria and viruses); the majority enter the intact plant surface and spread either intercellularly or intracellularly. There is seldom a general dissolution of cell walls beneath an infecting hypha, as might be expected if diffusible enzymes were involved. The plant cuticle may become thinner at a point of penetration and underlying wall microfibrils may lose their integrity. On the other hand, electron micrographs show little evidence of distortion in the wall, as might be expected if mechanical pressure alone were responsible. In general, fungi appear to employ a combination of enzymes and mechanical forces to facilitate penetration of plant cell walls. Following penetration three main types of host–parasite relationship may be identified: an unspecialised necrotrophic relationship with immature or senescent tissue, a specialised necrotrophic relationship and a biotrophic relationship.

These relationships have been studied at the physical, physiological and molecular levels. Physical results of penetration and colonisation are seen some time after invasion when the host–pathogen relationship is well established as the gross symptoms typical of a diseased plant (section 1.3). At this stage, particularly with necrotrophic relationships, host tissues are degraded possibly because of the effects of enzymes and toxins which may emanate from either the parasite or, most probably, the host (Husain and Kelman, 1959; Wood, 1967). Prior to this, events of a microscopic nature may occur as in vascular wilt diseases where there is a development of gels, gums and tyloses in the vascular system. In biotrophic relationships *haustoria* will be evident. These are specialised absorbing organs, varying in shape from knob-like through elongated to branched, which are formed by the pathogen through pores punctured in the host cell wall. When a haustorium penetrates a host cell it does not puncture the plasma membrane but invaginates around it. A function of haustoria is, therefore, to increase the absorptive area available to a parasite. It is considered that the key to the movement of solutes from host to pathogen, in powdery mildews at least, lies in those processes occurring at the interface between haustorium and host cytoplasm (Bushnell and Gay, 1978). At the ultra-structural level there are changes to cell-wall structure, cellular membranes and organelles during pathogenesis.

Much pathological investigation has centred on identifying changes at the physiological level with host–pathogen relationships, particularly with fungi. General effects include changes to host cell permeability, stomatal function, water absorption and conduction, respiration, photosynthesis, nucleic acid, protein and phenolic metabolism and growth regulator levels (Horsfall and Dimond, 1959; Wheeler, H., 1975; Horsfall and Cowling, 1978). Many of these responses are of a gross nature and similar to those found in plants damaged by non-microbial agents such as pollutants. Attention is now turning to the molecular level in an attempt to find the primary trigger that determines whether there will be a susceptible or resistant response to penetration by a pathogen. The change from healthy to diseased state may depend on interference with host DNA transcription or RNA translation.

2.5 FUNGAL CLASSIFICATION

Nomenclature and classification, which are reliable and reproducible, are essential tools for the identification and discussion of organisms whether they are hosts or parasites. Since the categories are constructed for convenience by man, irregularities and exceptions are inevitably common. An ideal classification should rely mainly on sexual characters but in the fungi this is often far from possible, consequently several groupings are unnatural. It is not possible to provide more than the broadest of generalisations concerning fungal classification; readers wishing for greater detail should initially consult Talbot (1971), Ainsworth *et al.* (1973), Alexopoulos and Mims (1979) and Webster (1980).

Fungi are divided into the following two divisions.

Myxomycota

These are non-mycelial, wall-less forms possessing plasmodia or pseudoplasmodia; the former are naked masses of multinucleate protoplasm and the latter are aggregations of separate amoeboid cells. Within the Myxomycota only the class Plasmodiophoromycetes contains important plant pathogens (section 2.6.1).

Eumycota

These are the true walled fungi, which either produce a filamentous thallus (mycelium) or are unicellular. They possess common features of organisation, nutrition, physiology and reproduction. Five subdivisions may be recognised within the Eumycota and these, together with those classes containing significant plant pathogens, are listed in table 2.1. This classification is based on a wide variety of sexual, asexual and structural criteria, some of which are discussed in section 2.6.

2.6 EXAMPLES OF PLANT-PATHOGENIC FUNGI

2.6.1 *Plasmodiophora brassicae*—the cause of clubroot in crucifers

Only a few plant pathogens are found within the Myxomycota but they are of major significance to crop production; for example, *Plasmodiophora brassicae* (clubroot of crucifers), *Spongospora subterranea* (powdery scab of potatoes) and *S. subterranea* f. sp. *nasturtii* (crook root of watercress).

Table 2.1

Outline of the subdivisions and classes within Eumycota

Subdivisions	*Classes (containing significant plant pathogens)*	*Examples (used in subsequent sections)*
Mastigomycotina	Oomycetes	*Phytophthora infestans* (2.6.2)
		Bremia lactucae (2.6.3)
Zygomycotina	Not of plant-pathological significance	
Ascomycotina	Hemiascomycetes	None
	Discomycetes	*Sclerotinia fuckeliana* (2.6.4)
	Loculoascomycetes	*Venturia inaequalis* (2.6.5)
	Plectomycetes	*Sphaerotheca pannosa* (2.6.6)
	Pyrenomycetes	None
Basidiomycotina	Hymenomycetes	*Armillaria mellea* (2.6.7)
	Teliomycetes	*Cronartium ribicola* (2.6.8)
Deuteromycotina	Hyphomycetes	*Verticillium* spp. (2.6.9)
	Coelomycetes	None

They are characterised by the production of plasmodia and of zoospores which have two unequal flagella (whip-like processes permitting movement). *Plasmodiophora brassicae* is an example of a gall-causing pathogen. Symptoms are found only on members of the Cruciferae, but *P. brassicae* is capable of infecting the root hairs of non-cruciferous hosts such as *Agrostis alba* var. *stolonifera* (fiorin), *Dactylis glomerata* (cock's foot), *Holcus lanatus* (Yorkshire fog), *Lolium perenne* (rye-grass), *Fragaria* spp. (strawberry), *Papaver rhoeas* (field poppy) and *Rumex* spp. (dock and sorrel).

Symptoms

Typical symptoms caused by *P. brassicae* are galling and contusion of the host root and hypocotyl. The extent of galling varies with the age, size and type of roots. Galls on swollen hypocotyls of crops such as swede (*Brassica napus*) and turnip (*B. campestris*) may partially or completely cover the swollen 'root' surface. The fibrous roots of cole crops (*B. oleracea*) become covered in coralloid galls. Prior to obvious gall symptoms, the host foliage becomes flaccid and blue-green indicating that host water metabolism is seriously impeded. Growth is retarded leading to stunted and unthrifty plants. Infection of cabbage (*B. oleracea* var *capitata*) alters the distribution of dry matter within the host. After clubbing symptoms become visible, increases in root growth are almost entirely attributable to growth by the galls, while growth of the foliage is diminished. Diseased plants tend to form fewer and smaller leaves which expand more slowly than those of uninfected plants. The extent of galling is dependent mainly on the period of time over which the plant is subject to infection rather than its age at first infection.

The effect of environmental factors on the severity of clubroot symptoms is not well defined. Disease severity will increase when soil moisture content is raised from 50 per cent of maximum water-holding capacity up to saturation. Temperature will also influence the number of plants showing symptoms; complete infection occurs only where daily mean temperature is above 19.5°C. Increases above this to 26°C will influence the severity of clubbing. Soil acidity is perhaps the most vexed factor associated with infection by *P. brassicae*. Although the disease is primarily associated with acid soils (pH less than 7.0) it is apparent that in alkaline soils (pH greater than 7.0) club development will take place provided that soil moisture content, temperature and spore load are sufficiently high. Infection is not inhibited by alkaline soils but the range of conditions suitable for infection is more limited than those for acidic soils. High light intensity 14 to 21 days after root hair infection (figure 2.3) indirectly influences the severity of symptoms by encouraging host growth. Spore concentration will influence infection but only limited investigations have been made which show that soils containing less than 1000 spores g^{-1} are unlikely to be associated with heavy clubbing. Work with nitrogen fertilisation is conflicting; some reports suggest that additional nitrogen will increase clubroot whereas others suggest the reverse.

Biology

The life cycle of *P. brassicae* is summarised in figure 2.3 (see also Tommerup and Ingram, 1971). The resting spore is an extremely efficient survival mechanism permitting this organism to remain viable for several years in the absence of a host. In the presence of a suitable host a spore germinates to release a single zoospore with two flagella of unequal length. The zoospores are vulnerable, thin-walled structures which invade the delicate host root hairs within 30 h of being released from the resting spore. During the invasion process the zoospore becomes encysted against the root hair adhering to it and eventually the zoospore contents are injected into the root hair (Aist and Williams, 1971). Inside the root hair cell, *P. brassicae* forms a spherical amoeboid body with close connection between parasite and host. By mitosis and cytoplasmic cleavage multinucleate zoosporangia form. The zoospores may escape from the originally invaded host cell to enter other root cells and colonise more deeply into the root tissues; it is at this stage that gall formation takes place. Secondary plasmodia form which, like the primary plasmodia, are immersed in host cytoplasm and surrounded by only a thin envelope. There is rapid host-cell extension and multiplication with high concentrations of hormones present in infected roots giving rise to the host-root symptoms of clubroot. The plasmodia

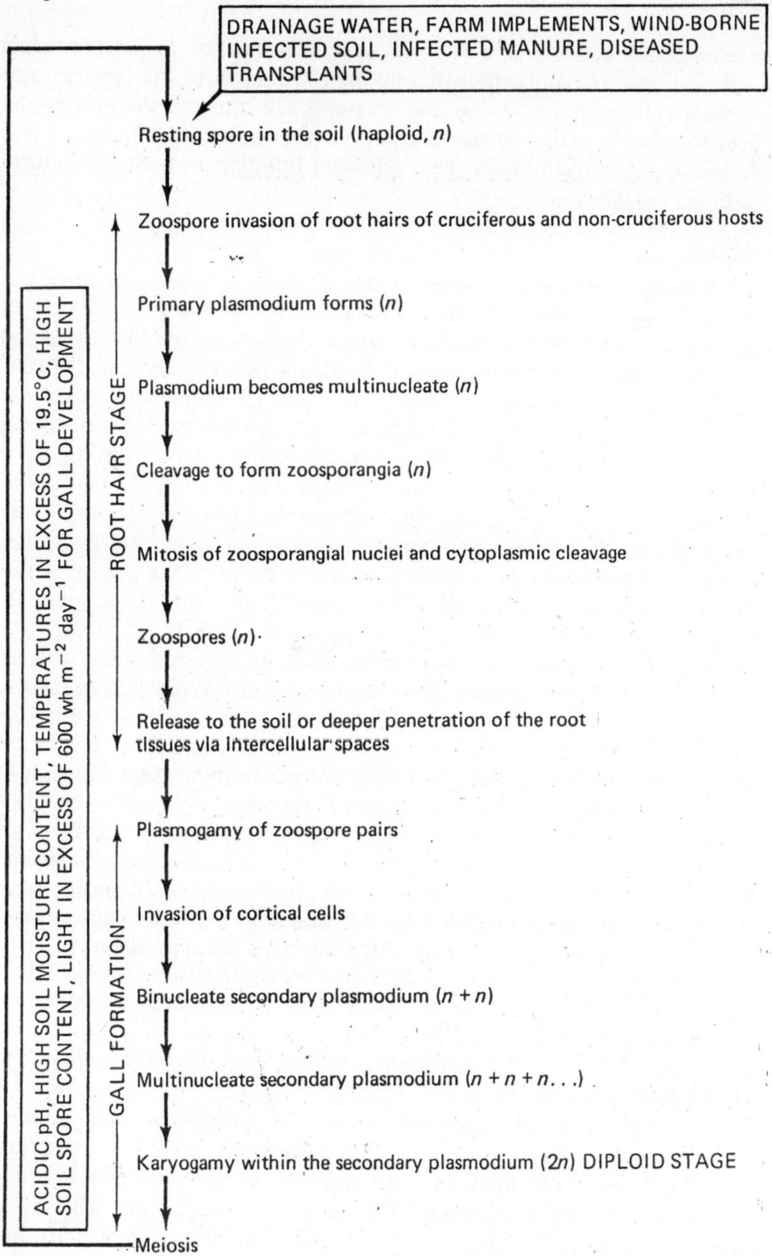

Figure 2.3 Infection cycle of *Plasmodiophora brassicae*—the cause of clubroot of crucifers

become multinucleate and it is thought that nuclear fusion producing diploid nuclei is rapidly followed by meiosis leading to the formation of haploid resting spores. These are released back into the soil in a passive process brought about by the decay of host roots. It should be noted that *P. brassicae* is a highly biotrophic organism fulfilling many of the features outlined in table 1.1.

Control

Control of *P. brassicae* has been achieved by use of resistant cultivars in turnip (*B. campestris*) and swede (*B. napus*) and these are used widely in agriculture. Horticultural brassica crops are largely of the fibrous-rooted cole type where the introduction of resistance has proved more difficult. Some resistance is thought to be present in kale (*B. oleracea* var. *acephala*) and the first commercially available cabbage resistant to *P. brassicae* was Badger Shipper developed at Wisconsin University, USA, from a chance cross between cabbage and kale. Attempts have been made to transfer resistance into Brussels sprout (*B. oleracea* var. *gemmifera*), cauliflower and broccoli (*B. oleracea* var. *botrytis*) but without great success so far. A major difficulty is to achieve a resistant cultivar that also possesses desirable agronomic characters. Resistance is also known to exist in some culinary radish (*Raphanus sativus*) cultivars. Genetic variation within the pathogen enables the development of populations that are able to overcome the host resistance introduced by plant breeders, consequently resistance mechanisms need to be broadly based in order to have lasting value. There is little or no information as to the importance of the unspecific root hair stage compared with the gall-forming stage and sites at which host resistance is operative against *P. brassicae*.

Chemical control has been largely directed at protecting the root systems of newly transplanted crops so that they become established before severe damage from gall formation has occurred. The major technique, which has been practised for decades, is to use 4 per cent calomel (mercurous chloride) slurry into which the roots are dipped just prior to transplanting. This technique protects the treated roots but gives no protection to those produced subsequently. Benomyl and related compounds may be used as less hazardous substitutes which also have slightly systemic properties. There is considerable interest in the use of chemical sterilants to eradicate *P. brassicae* from soil. Materials found to be effective include chloropicrin, dazomet, dichloropropane-dichloropropene and methyl bromide. Pentachloronitrobenzene (Quintozene) is toxic to *P. brassicae* but can, however, be phytotoxic to brassicas; its use with hydrated lime can ameliorate this disadvantage. Control of *P. brassicae* was achieved in the USA by using 5 tonnes ha^{-1} hydrated lime broadcast plus 10–20 g PCNB m^{-2} applied as a band to the crop rows. Application of lime to raise pH above 7.0 is a traditionally advocated method of clubroot control, but very heavy rates are needed; in Newfoundland and California, USA applications in excess of 45 tonnes ha^{-1} are reported. Applications of calcium cyanamide also reduce clubroot infestation.

Husbandry techniques remain a major means of clubroot control. Wide rotations, with at least a six-year break between any form of cruciferous crop, are essential. This applies with equal force to crops where resistant cultivars are used since it is likely that the root hairs of such cultivars are invaded during the first stages of infection. Even wider rotations could be advocated for spring-growing and summer-growing brassica crops which occupy the soil in those climatic periods that provide optimal conditions for *P. brassicae* activity. Soil cultivation can be used effectively for clubroot control; measures such as subsoiling which improve drainage will ameliorate the level of infection. In south-west Wales continual rotary cultivation of infected land intended for cauliflower production lowers the level of infestation. This is presumably because loose, dry, friable soil permits desiccation of the resting spores. Solar heating of soil is also known to reduce the levels of *P. brassicae* in soil (section 9.1.1).

2.6.2 *Phytophthora infestans*—a cause of blight in potato and tomato

Many of the most damaging plant pathogens in the subdivision Mastigomycotina are classed as Oomycetes. There are two families of prime importance, Pythiaceae and Peronosporaceae (section 2.6.3). The Pythiaceae contain *Pythium* and *Phytophthora* spp. which cause root and foot rot diseases. Some *Phytophthora* spp. cause foliar and fruit blights and *Pythium* spp. are often associated with damping-off syndromes. Some members of these genera are capable of existing as facultative parasites or saprophytes displaying both necrotrophic and biotrophic characteristics.

Phytophthora infestans is one of the most widely known plant pathogens, largely because of the devastating social effects following destruction of the Irish potato crops in the period 1845–50. The Irish had been forced to rely heavily on the potato as a staple food source and successive crop failures caused by *P. infestans* reduced the population by death and emigration by 20 per cent.

Symptoms

Several sources lead to the initial infection of a potato crop. These include infected 'ground-keeper' tubers which remain as weeds from previous crops, infected seed potatoes and sporangia blown by the wind from neighbouring crops already infected or from residues remaining at clamp sites. The first signs of *P. infestans* infection on potato leaflets, rachis, petiole or stem are the development of brownish to purple-black lesions of varying sizes which enlarge rapidly, spreading over the entire host plant in favourable weather. Each lesion has an outer purplish zone merging into the remaining healthy uninfected green tissue. Infected areas become water-soaked and soggy, rapidly reducing the entire plant to a blackened rotting pulp in a few days—a typical blight-disease syndrome. In dry weather foliar lesions remain as small dry brown patches. On the under surface of foliage opposite each lesion, areas of white fungal fructification (*sporangiophores* carrying *sporangia*) protrude through the stomata. Sporangiophores are formed densely in moist conditions and may be almost absent in dry weather. Developing potato tubers are infected by sporangia that are washed off the foliage and into the soil. In slight attacks, infected tubers show a dry rot that fails to soften the tissues but produces rusty-brown markings just below the surface, which may extend in a variable pattern into the tuber. In wet conditions a vigorous dark smelly rot develops throughout the tuber; much of this rotting is caused by secondary organisms.

Symptoms on tomato foliage are generally less severe than, but usually similar to, those found on potato. Tomato fruits are particularly vulnerable to infection by *P. infestans*. Invasion can take place at any stage in fruit development, usually beginning at the stem end. Large shapeless grey to greenish water-soaked lesions form on semi-mature and ripe fruits. Infection of green fruits leads to lesions that are dark brown with a firm corrugated texture, possessing a definite border of green uninfected tissue around them. Under conditions of high relative humidity, sporulation may be seen on the fruit.

Biology

Phytophthora infestans reproduces sexually via oogonia and antheridia; resultant plasmogamy and karyogamy produce oospores. Meiosis is thought to occur during gametangial development. Oospores form only when isolates of different types of *P. infestans* are paired together. Isolates are of two compatibility types, A_1 and A_2, controlled by two alleles at a single locus (Savage *et al*., 1968). This process of self-incompatibility is called *heterothallism*. Both compatibility types occur together only in Mexico and hence sexual reproduction is restricted geographically (figure 2.4); this area is thought to be the site of origin of both *P. infestans* and the potato. Elsewhere reproduction is entirely asexual. Infection establishes an intercellular mycelium from which curled haustoria are produced into host cells. Sporangiophores form from the mycelium through stomata and bear abundant lemon-shaped sporangia. These germinate either indirectly forming motile zoospores or directly producing a germ tube. Temperature determines the form that this process takes; below 15°C zoospores form, above 20°C germ tubes develop.

Environmental conditions greatly influence the life cycle of, and disease development caused by *P. infestans*. Sporangiophores do not form at relative humidities of less than 85 per cent and sporangia require more

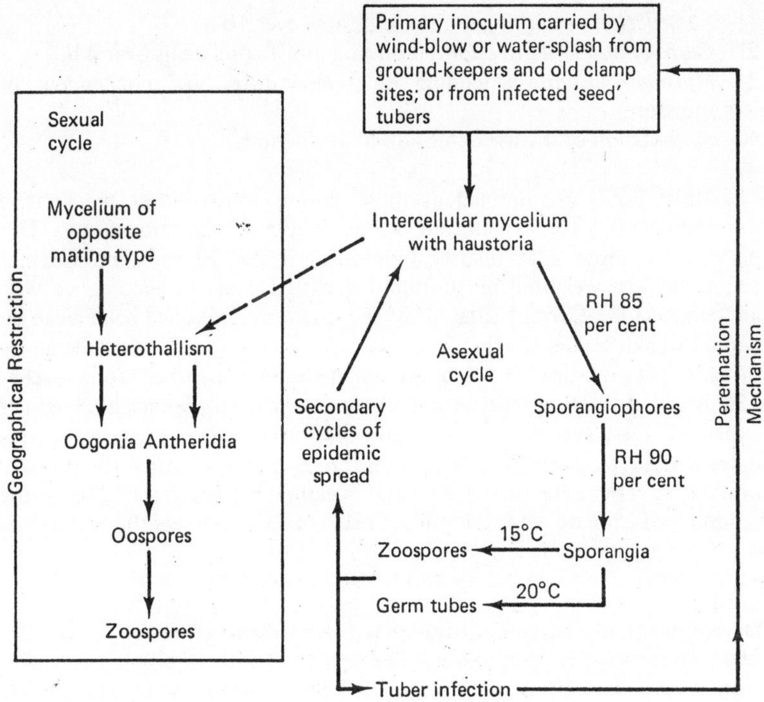

Figure 2.4 Infection cycle of *Phytophthora infestans*—the cause of late blight of potato and tomato

than 90 per cent relative humidity to form. Consequently weather plays an important role in the development of epidemics and knowledge of the interrelation of fungus and environment can be used to predict when crops are most at risk from *P. infestans*. Work by Beaumont and Staniland (1933) in South Devon showed that where temperature was above 10°C for 48 h and relative humidity greater than 75 per cent, blight would develop on potato crops within 10 days. Such periods of favourable conditions for *P. infestans* development are known as 'Beaumont Periods'. Subsequent modifications have been made and several prediction equations applied; they all underline the necessity for a period of 15 h of free water at the host surface during which the sporangia germinate and invasion takes place. One of the most sophisticated systems is the West German 'Phytoprog' (Burckhardt and Freitag, 1969). This takes into account four different stages in the fungus life cycle related to type of weather.

1. *Sporulation*—requiring moist conditions over 10 h.
2. *Germination and infection*—requiring moist conditions over 4 h.
3. *Mycelial growth*—depending on temperature but independent of moisture.
4. *Suppression* of disease extension in dry periods.

Moist hours are defined as those during which either the relative humidity is at least 90 per cent or measurable precipitation occurs. Dry periods are those with relative humidities below 70 per cent. Weather sequences are weighted by multiplying with factors in accordance with air temperature. Current total weather ratings are reckoned each week by adding weekly values starting from a known average date for the emergence of early potato crops. The date on which the total weather rating reaches 150 is called the first 'critical date' before which no appreciable blight is expected. First symptoms are usually seen 10–40 days later and so this date is used to alert growers to keep crops under scrutiny for primary infections. The date on which total weather ratings reach 270 is the second 'critical date' and when this is reached a call is issued for immediate control measures to be applied.

Control

Development of resistant cultivars has been a major objective for control of *P. infestans*. For this, use has been made of interspecific crosses with wild species notably *S. demissum* from South America. Initially it was hoped that interspecific crosses would result in resistance that operated by sealing off the invading pathogen by death of the host cells immediately in front of the advancing mycelium (*hypersensitive resistance*). It was soon apparent, however, that despite an apparently solely asexual reproductive cycle in Europe and North America, *P. infestans* populations contained physiological races capable of circumventing such resistance mechanisms. A series of races has been identified that are compatible with the eleven resistance (R) genes and various combinations of them (Malcolmson, 1969). More recently, plant breeders have looked for resistance related to preventing the initial entrance or slowing the rate of subsequent growth of *P. infestans* in leaves and tubers.

Once infection is established in a crop, chemical control is the sole means of checking further spread. Chemicals are applied as sprays, or less frequently as dusts, when less than 0.1 per cent of plants are infected and hence application is timed using forecasting systems. Various chemicals may be used; Bordeaux and to a lesser extent Burgundy mixtures still provide effective control but they are being replaced by dithiocarbamate materials such as mancozeb, zineb and maneb and organo tin materials such as fentin hydroxide and fentin acetate. At the end of the growing season haulm destruction is essential to help prevent tuber invasion and also to facilitate easier mechanical harvesting. Sulphuric acid is used widely as a haulm desiccant but other materials such as dinoseb or diquat are available.

Husbandry measures are also important for *P. infestans* control; these include effective earthing-up which helps to prevent sporangia from reaching the tubers. Sporangia may survive in soil for several weeks but they do not penetrate deeply into it. Removal of ground-keepers can be achieved by crop rotation with cereals where hormonal herbicides can be used to control broad-leaved 'weeds'. At clamp sites and storage areas, debris that may be infected with *P. infestans* should be destroyed.

2.6.3 *Bremia lactucae*—downy mildew of lettuce

The downy mildews affect a wide range of horticultural and other crops such as brassicas, beet and spinach, cucurbits, onions, peas and beans, hops, vines, ornamentals and soya beans and there is a steadily increasing appreciation of their effects on graminaceous hosts. Only infrequently are host plants killed by downy mildew pathogens; usually a biotrophic relationship exists which leads to disfigurement but nonetheless considerable crop losses. Downy mildew fungi and *Phytophthora* spp. (section 2.6.2) are closely related, being classed together as Oomycetes. *Phytophthora* spp. are facultative parasites and placed in the family Pythiaceae, whereas downy mildews are obligate parasites and placed in the family Peronosporaceae. Although *B. lactucae* infects several members of the Compositae, attempts to cross-inoculate isolates obtained from one host species to another have been largely unsuccessful. Isolates from field and protected lettuce crops may only be reinoculated back on to lettuce cultivars and a few wild *Lactuca* spp.; *B. lactucae* is highly host specific.

Symptoms

Lettuces at all growth stages from seedlings to mature plants are susceptible to *B. lactucae*. Usually the first signs of infection are the appearance of sporophores growing through the stomata from an intercellular mycelium established within the host tissues. These discrete sporangiophores bear sporangia and are most frequently seen on the undersurfaces of leaves. In severe infections of mature plants and more generally with seedling infections, an entire plant may be covered in sporulations. Seedlings may be killed by such infection. Sporulation is accompanied in maturing plants by a development of light-green to yellowish angular lesions on the upper surfaces of leaves opposite to sporulations on the undersurfaces. Lesion growth is limited by the larger leaf veins and this results in the angular shape of lesions. When seed-bearing plants are infected there is invasion of flower stalks so seed production and quality are impaired. Systemic infection causes a black-brown discolouration of stem tissues and leaf bases near the shoot tips of mature heads.

Biology

Invasion most frequently occurs by direct penetration through the epidermal cells walls although entry may take place through stomata. After penetration of an epidermal cell *B. lactucae* forms intercellular hyphae which branch and ramify throughout the host tissue. As is characteristic of biotrophic pathogens, haustoria are produced by *B. lactucae*; these form at intervals from each intercellular hypha, penetrating and developing within adjacent host cells. The formation of oospores as a result of sexual reproduction has been recognised in *B. lactucae* for some time but only recently have the distribution and importance of oospores as mechanisms for variation and perennation been appreciated. Heterothallism, with the existence of two complementary strains of *B. lactucae*, B_1 and B_2, was described by Michelmore and Ingram (1980). These authors also found homothallic strains of *B. lactucae*. Possession of both mechanisms possibly gives some explanation for the high level of variation, particularly for pathogenicity, found in *B. lactucae*, Asexual reproduction takes place within 5-14 days of colonisation; copious sporangia develop providing a means of epidemic spread aided by agencies such as wind and water-splash. The sporangia of most Peronosporaceae are able to germinate indirectly by producing zoospores (for example *Peronospora parasitica* and *P. viciae*; in contrast those of *B. lactucae* germinate directly by germ-tube formation. The requirements for growth and sporulation by *B. lactucae* may be summarised as a cool moist environment as outlined in figure 2.5. Transmission is not thought to take place by seed or from wild host species but mostly via sporangia carried from one crop to another. It is now realised that oospores present in lettuce debris provide an important mechanism for perennation from one crop to the next.

Control

There have been continual attempts to control *B. lactucae* by breeding resistant lettuce cvs over the last 50 years. These have rested on use of major genes (chapter 5) and have resulted in more than 100 cvs possessing race-specific resistance. Concurrently at least 30 physiological races of *B. lactucae* have been identified in Western Europe, Israel and the USA. Nomenclature of these races solely in relation to the susceptibility and resistance of cvs in these centres of crop production resulted in a confused picture. This was clarified by a genetic interpretation which identified five dominant resistance genes and a pair of genes with complementary effect in the host and postulated at least ten matching virulence factors in the pathogen (Johnson *et al.*, 1977). A survey of virulence-factor frequency showed them to be widespread in *B. lactucae* populations (Dixon and Wright, 1978) and that further breeding effort using currently available major genes was unlikely to be cost-effective. Development of resistant

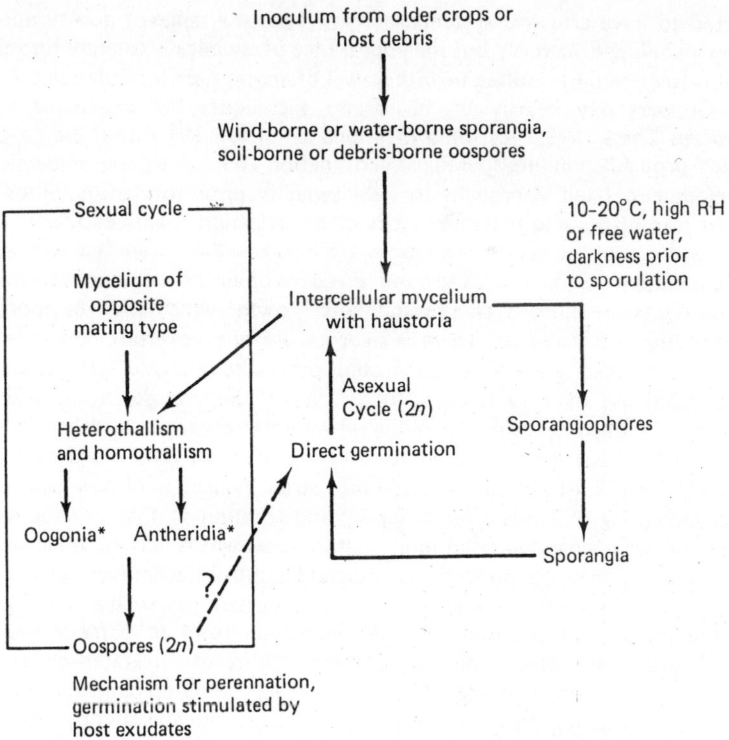

*Meiosis is thought to occur within the oogonia and antheridia; for most parts of the life cycle *B. lactucae* is diploid

Figure 2.5 Infection cycle of *Bremia lactucae*—downy mildew of lettuce

cvs is more likely to succeed by incorporation of genes from wild relatives of lettuce, such as *Lactuca saligna* and *L. virosa*, or by use of non-specific resistance. The latter will, however, require an acceptance by the market of some level of infection on lettuce heads.

Effective control of *B. lactucae* by chemical means requires frequent applications of large quantities of material. Initially copper-based and sulphur fungicides were used. Dithiocarbamate compounds such as mancozeb, thiram and zineb have replaced them. Dust formulations are particularly advocated because their use does not result in increased RH within the crop. Applications are made every seven days or more frequently if there is a nearby source of inoculum. Phytotoxic effects may develop in protected crops following the use of dust formulations if night temperatures fall below 6°C. Recently, derivatives of acylalanine have been demonstrated to have effective systemic activity against a range of downy mildews including *B. lactucae* but the appearance of chemically tolerant fungal strains (section 6.6) resulted in withdrawal of appropriate formulations.

Growers rely heavily on husbandry techniques for control of *B. lactucae*. These largely rest on a maintenance of low RH within the crop which probably inhibits sporangial germination and consequent epidemic development. Land is brought to field capacity prior to planting block-raised transplants and thereafter little or no irrigation is applied. Particularly with protected crops, the blocks are stood on the soil surface so that air circulation is encouraged around the developing plants. Applications of nutrients are limited to prevent 'soft growth' which may be more susceptible to *B. lactucae*. There is rigorous use of preplanting herbicides to prevent weed growth which would otherwise increase RH. Under protection, use of a 12°C day and 5°C night temperature regime with pipe heat applied to dispel humidity in mid-afternoon was widely advocated in the Netherlands for winter and spring crops. Crop management to avoid infection by *B. lactucae* requires strict hygiene to remove debris, deep ploughing of fields after cropping and rotation so that developing crops are well separated from more mature ones which may be infected. Eventually it may be possible to integrate control techniques (section 9.1.6) since it has been shown that some field-resistant cvs, such as Iceberg, require lower levels of fungicide for protection from *B. lactucae* than highly susceptible ones. This raises the possiblity of integrated genetic and chemical control providing a more lasting means of control which could not be achieved by either technique alone (Crute and Dixon, 1981).

2.6.4 *Sclerotinia fuckeliana* (stat. conid. *Botrytis cinerea*)—the cause of grey-mould rotting and dieback of many crops

Members of the subdivision Ascomycotina such as *S. fuckeliana*, *Venturia inaequalis* (apple scab—section 2.6.5) and *Sphaerotheca pannosa* (rose powdery mildew—section 2.6.6) are largely classified by their sexual stage. This is a sac-like ascus typically containing eight ascospores. These may develop within a protective structure (*ascocarp*), morphological characters of which are used to describe classes within the Ascomycotina. Characteristics of some classes are as follows.

Hemiascomycetes

Members of this class may be grouped into those where a mycelium is absent and each vegetative cell acts as an ascus—they are mainly saprophytic types such as yeasts (*Saccharomyces* spp.); and those that form asci from terminal cells of the mycelium. The latter usually produce a continuous layer of spore-forming cells interspersed by occasional barren hair-like cells (*paraphyses*); such a structure is termed a *hymenium* (figure 2.6). Examples include *Taphrina* spp. which cause leaf curl, leaf blister and witches' broom, particularly parasitising members of the Rosaceae.

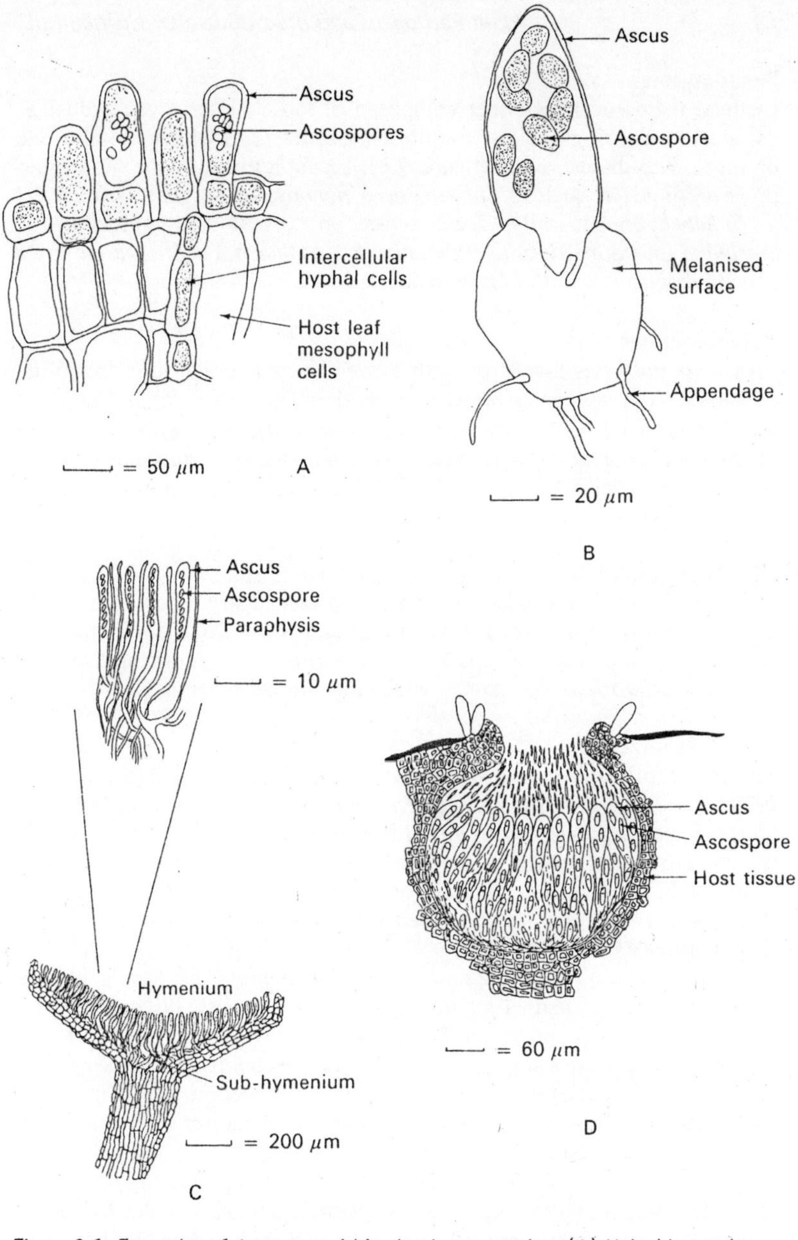

Figure 2.6 Examples of Ascocarp within the Ascomycotina. (A) Naked hymenium: section of host leaf showing hymenium of *Taphrina deformans* (peach leaf curl) (after, Hawker, 1967). (B) Cleistothecium: from *Sphaerotheca pannosa* (rose powdery mildew) (after Webster, 1980). (C) Apothecium: from, for example *Sclerotinia* spp. (after Alexopoulos and Mims, 1979). (D) Pseudothecium: from *Venturia inaequalis* (apple scab) (after Sivanesan and Waller, 1974)

Plectomycetes

In these the ascocarp is either composed of loosely interwoven hyphae or compacted into a globular structure (*cleistothecium*) which contains one or more club-shaped ascus (figure 2.6). Examples include the biotrophic powdery mildews such as *Sphaerotheca pannosa* on rose (section 2.6.6), *S. fuliginea* on cucurbits, *S. mors-uvae* on currants and gooseberry, *S. macularis* on raspberry and strawberry, *Podosphaera leucotricha* on apple and *Microsphaera alphitoides* on oak.

Pyrenomycetes

This class possesses ascocarps with a sac-like structure around the inside of which is a hymenial layer composed of single-walled (unitunicate) asci interspersed with paraphyses (*perithecium*). The whole structure opens by a pore or *ostiole*. Perithecia occur singly or clustered together within a mycelial matrix. Examples include *Ceratocystis ulmi*—the cause of Dutch elm disease.

Discomycetes

Here the ascocarp is a saucer-shaped or cup-shaped structure within which a hymenium is freely exposed when ripe and is known as an apothecium (figure 2.6). Members of this class are termed 'cup fungi' and include species of *Sclerotinia* that cause a wide range of soft rot diseases.

Loculoascomycetes

The ascocarp in this class usually, but not invariably, resembles a sac-like perithecium and is termed a *pseudothecium* (figure 2.6). These form from a mycelial matrix interspersed within which are several bitunicate asci; as growth proceeds the whole structure becomes one to several locular. It is the unilocular forms that have ascocarps most resembling perithecia. Examples of this class include *Venturia inaequalis*—the cause of apple scab (section 2.6.5).

The sexual and asexual phases in the life cycle of a fungus may have been described and named separately before mycologists discovered them to belong to a single organism. In such cases the sexual (*perfect* or *teleomorph*) stage provides the 'proper' name but frequently the name applied to the asexual (*imperfect* or *anamorph*) stage has become widely used and recognised. When eventually the complete life cycle has been elucidated the asexual state name may be retained as a suffix and termed 'status conidium' (stat. conid.). This is the situation with *Botrytis cinerea*, where the perfect stage has been recognised as *Sclerotinia fuckeliana* but *B. cinerea* is retained when discussing attributes of the conidial state. Additionally there are species of *Botrytis*, many of which are important crop pathogens, for which no perfect state has yet been identified. Pathologically *Botrytis* spp. may be divided into two groups; the aggregate species *B. cinerea* which infects a vast host range and other species which have relatively more specialised parasitism, such as *B. allii*, *B. byssoidea* and *B. squamosa* the cause of rots of *Allium* spp., *B. tulipae* the cause of tulip fire and *B. fabae* the cause of chocolate spot on *Vicia* beans.

Symptoms

Botrytis cinerea is possibly the most extensive pathogen of horticultural crops, causing very considerable crop losses. Paradoxically *B. cinerea* is weakly pathogenic in its ability to penetrate tissue, usually invading through sites of injury, mechanical damage or via thin-walled tissue such as floral organs or developing leaves or shoots. The profuse production of superficial off-white to grey mycelium makes it an easily identified pathogen. The mycelium sporulates profusely from dark-coloured conidiophores. Once this fungus is established, infected tissue is rapidly destroyed and reduced to a water-soaked mass of slimy tissue covered in grey mould fructification. Water-soaking is due to the effects of cell-macerating enzymes produced by the pathogen. Many features of a typical necrotrophic pathogen are displayed by *B. cinerea*. Infections by *B. cinerea* can remain in a latent phase causing little more than small circular foliage lesions with scant sporulation when environmental conditions are adverse to pathogen growth and propagation. Once the environment is conducive, a rapid growth phase ensues, the aggressive stage, which produces abundant conidia, by which epidemics can develop very rapidly. Examples of crops that are attacked include *Phaseolus* beans and peas where pods are a prime site of invasion especially if they are in contact with the soil; similarly Brussels sprout buds, cabbage hearts and cauliflower curds become covered in sporulating grey mycelium following periods of heavy rain or frost damage; root crops such as carrots and parsnips are reduced to moribund masses of grey mould when stored under adverse conditions especially if injury occurs during lifting. Major losses due to *B. cinerea* are recorded in the lettuce crop because the leaves are composed of thin-walled parenchyma tissue that provides a ready site for penetration; lettuce plants may be invaded at any growth stage but the plant base and outer leaves, particularly of butterhead types in contact with soil, are often the first areas to become infected. Thereafter the heart is invaded becoming covered in conidia and turning to a water-soaked mass. Cucumber and tomato plants are frequently invaded via badly made pruning cuts from where *B. cinerea* may become systemic, moving in the vascular system to other parts of the host. A general soft rot of stems, leaves and fruits may develop especially in tomatoes where infection can additionally develop from dehisced flower calyces lodged on them and forming foci of infection. Tomato fruits may be disfigured by 'ghost spotting' where *B. cinerea* conidia have failed to invade the fruit but enzymes secreted during the abortive invasion process mark the fruit surface causing spots up to one cm diameter with a pale-green halo and silvery centre; such fruits are unsightly and lose quality grading. Soft fruits such as raspberry and strawberry can rapidly be reduced to a mass of grey sporulation by *B. cinerea*, particularly if the fruit is damaged in picking or wet at harvest. Flower crops such as *Chrysanthemum* and *Dahlia* can be ruined by *B. cinerea* infection which is often facilitated by maintenance of high RH in the glasshouse.

Biology

A generalised infection cycle for the sexual and asexual states of *S. fuckeliana* is shown in figure 2.7. In disease syndromes it is the asexual *B. cinerea* state that is predominant but this highly adaptable pathogen produces several propagules which are important in dispersal, perennation and reproduction. Conidia are the main agent for spread and epidemic development. They are dry spores dispersed largely in air currents and in some circumstances in or on water droplets. Experiments with the *B. cinerea* state in raspberry plantations showed that conidial release was related to changes in RH. Two major periods of release were identified; the first in the forenoon when RH was falling and the second in the

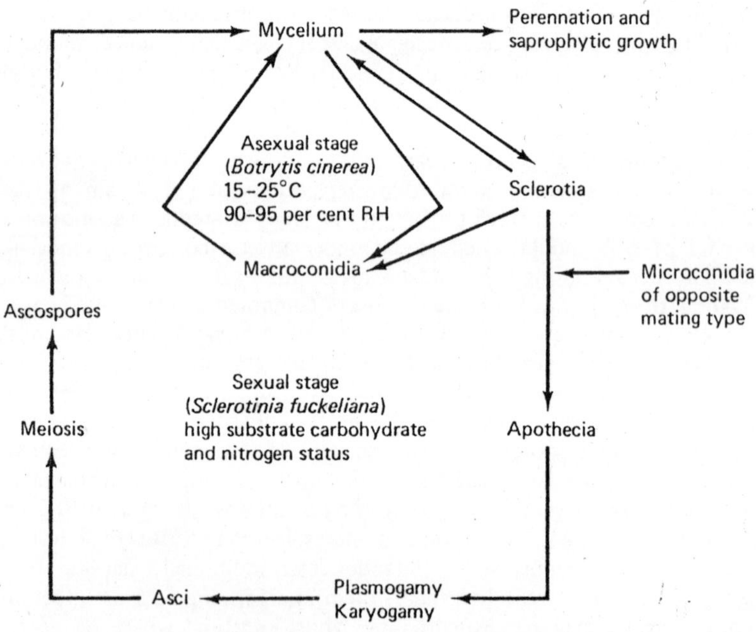

Figure 2.7 Infection cycle of *Sclerotinia fuckeliana (Botrytis cinerea)*—the cause of grey mould rotting and other diseases on a wide range of crops

afternoon when dew formed and RH rose. It was concluded that rapid change in RH but not the direction of the change results in conidial release. This pattern was modified by the effect of temperature during the preceding night. Temperatures below 12°C prevented conidial formation. Rainfall increased conidial release by creating turbulent air currents and conidia could be carried on water droplets themselves (Jarvis, 1980).

Mycelial fragments and possibly chlamydospores provide mechanisms for soil-borne and detritus-borne perennation with the additional possibility of dispersal by wind and rain-splash. Sclerotia are likely to be the main avenue of perennation and are well adapted to withstand fluctuating and extreme environmental conditions. Sclerotial initiation is favoured by high concentrations of carbohydrate and to a lesser extent nitrogen in the substrate (Townsend, 1957). Maturation of sclerotia does not occur until mycelial growth has been checked either by exhaustion of nutrient supply or by some metabolic change. Factors that favour sclerotial formation depress conidial production and vice versa. Sclerotia germinate to form conidia, mycelia or apothecia. For the latter, fertilisation by microconidia of the opposite mating type is required. Apothecial development leads to the formation of ascospores which may be infective spores giving rise to a mycelium and subsequent conidial production.

Penetration of a host from conidia or saprophytic mycelium may be accomplished by entry through the host cuticle either by mechanical or enzymic means, via stomata or, most frequently, through existing wounds. The most common syndrome that ensues is a necrotrophic soft rot brought about by enzymic degradation of the host cell walls. *Botrytis cinerea* has been shown to possess many of the enzymes necessary for host dissolution.

This pathogen is characterised by an ability to produce both latent and aggressive infections. The aggressive, rapidly spreading and destructive diseased state is mostly seen on older, particularly senescent, and very young tissues. Although the onset of an aggressive growth phase may be related to weather, it is also regulated by the physiological age of host tissue. Newly opened strawberry flowers are, for example, very susceptible to infection and the resultant blossom blight kills or damages many flowers. There may also be the establishment of a latent infection that displays no pathogenic effects of 30–35 days until the fruit is ripening. At this stage a devastating soft rot may develop (Jarvis and Borecka, 1968). Low infection levels of 1–5 per cent at the flowering stage are capable through latent infection producing losses of 10–20 per cent of strawberry fruits, depending on cultivar and plantation age.

Control

Chemical and cultural techniques provide the major means of controlling *B. cinerea*. Genetic variability within this fungus precludes breeding for resistance using traditional techniques with monogenic resistance. It may be possible, however, to accumulate polygenes and thereby to produce lines that exhibit reduced levels of disease expression.

Control by cultural techniques aims to reduce the levels of inoculum to which crops are exposed and to change the environment in order to favour the host rather than the pathogen. Strict crop hygiene, removal of dead and moribund plant tissues and careful husbandry whereby wounding is minimised reduce the inoculum potential of *S. fuckeliana*. Thus, for example, when pruning or picking tomatoes or cucumbers, cuts should be made cleanly without leaving broken snags which may be more easily invaded. The environment is manipulated to reduce RH levels. This may be achieved by low-density planting and rigorous use of herbicides for field crops while in protected crops measures are taken to encourage air circulation and to dispel water vapour within the crop. For most crops excessive use of nitrogenous fertilisers should be avoided in order to prevent the growth of thin-walled tissues that are more easily invaded. When harvesting crops, they should be handled carefully to ensure minimal damage. Root crops, for instance, should not be allowed to fall excessive distances from elevators into trailers or on to concrete floors (section 10.3.1).

Fungicides provide the most directly effective means of controlling this pathogen. Recently the main materials used have been systemic fungicides of the benomyl, carbendazim and thiophanate-methyl types. A successful spray programme should also include materials of a non-systemic type to prevent the accumulation of chemically resistant races of the pathogen population (section 6.6). Such materials include captan, chlorothalonil, dichlofluanid, dicloran, iprodione, tecnazine and thiram. Selection depends on the approval of individual chemicals for the crop in question, the period between spray application and harvest and cost effectiveness (chapter 6).

2.6.5 *Venturia inaequalis*—the cause of apple scab

Venturia inaequalis infects dessert, culinary and crab apples, ornamental *Malus* spp. and the loquat (*Eriobotrya japonica*). Severe damage is caused by reducing fruit quality and weakening the host through defoliation. This fungus is universally distributed in all apple-producing areas of the world and is considered one of the major pathogens of apples. A closely related species *V. pirina* causes similar damage to pears. A member of the Ascomycotina, *V. inaequalis* is classed as a Loculoascomycete since the perfect ascocarp stage is a pseudothecium; there is an important asexual stage (*Spilocaea pomi*) which produces numerous generations of conidia in summer and is responsible for the epidemic spread of this pathogen.

Symptoms

Foliar symptoms develop as scattered, roughly circular brown to olive-green spots surrounded by a finely branched margin. The margin is caused by *V. inaequalis* mycelium radiating from each lesion. Gradually the lesions turn grey and necrotic, possibly dropping out with a resulting shot-hole effect. Localised malformations may also develop as thickening and puckering of the leaf lamina around infected areas which have a brown velvety texture in moist weather due to the production of asexual conidia. Foliar symptoms vary in severity with different apple cultivars; sometimes the leaves are covered in irregular diffuse spots or may develop a blistered effect because of an inequality of growth below the lesions, or the leaf blade may even appear scorched. Lesions may arise on either surface of the lamina but on the youngest leaves as they emerge from the bud the exposed lower surface is first affected. In early season, before bud burst, infection may be detected as lesions on the bud scales and later on the flower sepals. Bud scale and shoot lesions are important elements in the perennation mechanisms of *V. inaequalis*. Lesions occur later on the sepals and flower stalks and in severe infections may girdle the fruit stalk leading to abscission of fruit or flowers.

Developing fruits are worst affected by apple scab and symptoms may develop at any growth stage. Initial symptoms on young fruits are the formation of small dark spots with a branched margin beneath the fruit cuticle. Such early lesions may lead to deformity of growth and often fruits will crack and remain immature. As the lesions develop and sporulate, infection spreads to neighbouring leaves and trees. On ripening fruit lesions are usually small, blackened or chocolate coloured with an unbroken but slightly abraded surface. Lesions may develop further during storage as small pitted or saucer-shaped brownish-black, black or dull-grey zones often concentrated near the stalk end of the fruit. The dendritic type of lesions typical of infection by *V. inaequalis* develop because, once a spore has landed and germinated, penetration of the cuticle ensues. This establishes a mycelial mass below the cuticle from which hyphae grow out to produce this branching pattern. From the subcuticular hyphae short brown conidia develop. Rupture of the cuticle follows, owing to pressure from fungal growth. Individual host cells show little evidence of injury until some time after invasion, often remaining intact for 3–4 weeks after infection. Thereafter the epidermal cells become increasingly vacuolated, turn brown and collapse; a cork cambium forms beneath the necrotic host tissue and, as a result of growth pressures from the fungal tissue, is pushed up to form a typical scab lesion. Such symptoms are seen particularly well on apple fruits where there is a stimulation in growth of pith cells in addition to cork formation. Conidia form only in areas of a lesion where there is no necrosis. When epidermal host cells die the fungal conidiophores cease sporulation and this leads to the gradual disappearance of the velvety covering of spores on the lesion.

Biology

The development cycle of *V. inaequalis* may be divided into two parts (figure 2.8). Primary infections develop in spring as a result of the release of ascospores from pseudothecia, giving rise to lesions which produce

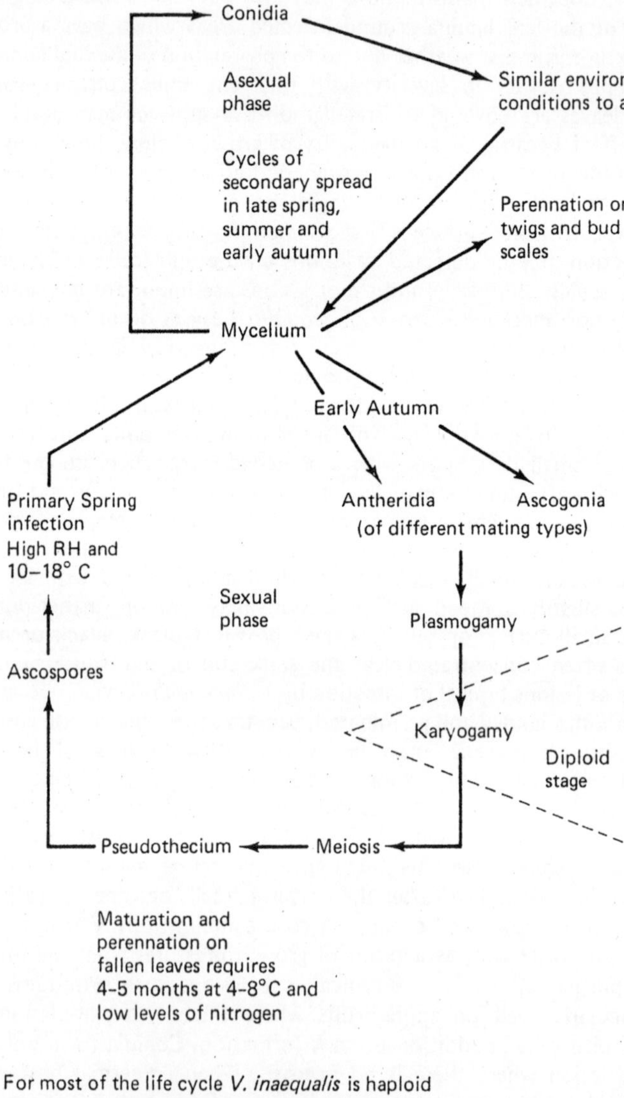

Figure 2.8 Infection cycle of *Venturia inaequalis*—the cause of apple scab

conidia. The latter are responsible for secondary cycles of infection on foliage and fruit from late spring to early autumn. The sexual cycle then recommences, maturation of pseudothecia being completed after leaf fall. Consequently this pathogen is capable of both parasitic and saprophytic modes of life. Development of pseudothecia requires the presence of compatible strains of *V. inaequalis* in the leaf, the fungus being heterothallic. Pseudothecia form over a period of 4–5 months at temperatures of 4–8°C and are encouraged by low levels of tissue nitrogen. Their numbers are dependent on the level of leaf infection in the previous autumn and survival rates during winter; the rate of pseudothecial maturation is related to host cultivar (Jeger *et al.*, 1982). Each pseudothecium contains several asci possessing typical two-celled ascospores which are of unequal size and give this species its name. Perennation may also occur by means of dormant mycelium present in twigs and bud scales but the importance of these as a source of inoculum varies with locality (Hirst *et al.*, 1955).

Ascospore release is largely related to the incidence of rainfall; dew formation leads to some liberation but is relatively unaffected by temperature. Release takes place over a period of 2-3 months; it is usually at a maximum in April–May in Britain. The subsequent infection process is also dependent on periods of leaf wetness. As with *Phytophthora infestans* (section 2.6.2) a relationship has been established between weather and severity of infection. This is quantified as 'Mills Periods' (Mills and La Plante, 1954). This relationship may be tabulated to show the number of hours that leaves need to remain wet, at different temperatures, for infection to take place if ascospores are present. For example

Mean temperature (°C)	5.6	7.2	10.0	11.7	15.0	15.6
Hours of leaf wetness for infection (h)	30.0	20.0	14.0	11.0	10.0	9.5

Ascospore germination is encouraged by temperatures of 10–18°C. Conidia, like ascospores, require cool, wet germination conditions but numbers of secondary infections may often be limited if control of primary ascospore infection is effective.

Control

Some apple cultivars are reported to possess levels of resistance to *V. inaequalis* but the strength of expression varies with climatic and geographical conditions. A single dominant gene for resistance was identified in the 1920s in Illinois, USA, from *Malus floribunda*. This resistance has been exploited in Canada, Europe and the USA together with that from other wild species such as *M. micromalus, M. atrosanguinea*, and *M. prunifolia*. Resistance in *M. atrosanguinea* has been related to a hypersensitive reaction following infection and associated with release of phloridsin by the penetrated tissues. More generally, resistance has been correlated with a number of factors, such as plant vigour, low organic acid and high catalase and peroxidase activities, leaf pubescence and the influence of the rootstock. Polygenic resistance is found in cv. Antonowka where resistance to all known races of *V. inaequalis* is claimed. Further sources of this resistance are found in selections of *M. baccata*, *M. sargenti*, *M. sieboldii* and *M. zumicalocarpa*. This resistance type has been utilised extensively in breeding programmes in Europe.

Three main strategies are applied to the use of fungicides and these are related to the life cycle of *V. inaequalis*.

1. *Post-infection or eradicant spraying* applied within 48 h of an infection (Mills) period to eradicate new infections.
2. *A protective schedule* applied at approximately 14-day intervals from bud burst to fruitlet stage irrespective of weather conditions.
3. *Post-harvest spraying* to disrupt the winter development of pseudothecia and spring production of ascospores.

These strategies are not rigid and may often be mixed to suit the circumstances of a particular producer. An extensive list of materials is available, for example, benomyl (P, S), captan (E, P), carbendazim (P, S), copper (P), ditalimfos (E, P), dithianon (P), dodine (E, P), mancozeb (P), manganese–zinc dithiocarbamate (P), mercury (E), propineb (P), sulphur (P) and thiophanate-methyl (P, S), where E = eradicant, P = protective and S = systemic action. Care must be taken in the use of some materials; for example, sulphur causes russetting to Cox's Orange Pippin fruit and ditalimfos cannot be used on Golden Delicious. Systemic materials such as benomyl, carbendazim and thiophanate-methyl have similar structure and properties. They can be applied less frequently than non-systemic chemicals but persistent use may lead to the development of fungicide-tolerant strains of *V. inaequalis* (section 6.6). Consequently they should be alternated with materials having a different mode of action.

After harvest, the foliage may be sprayed with phenyl mercuric materials to destroy developing pseudothecia. In the USA dinitro-orthocresol at 1 per cent in petroleum is used to destroy fallen leaf litter in spring, preventing primary ascospore infection. Natural destruction of pseudothecia in leaf litter occurs through the agency of earthworms. Breakdown of infected leaves may be accelerated by applications of nitrogeneous fertilisers such as ammonium sulphate or by spraying the trees before leaf fall with 2–5 per cent urea. This increases microbial activity and consequently leaf decay. Release of ascospores following these treatments was reduced to 3 per cent of that in unsprayed controls and pseudothecia failed to form in treated leaves.

2.6.6 *Sphaerotheca pannosa*—powdery mildew of rose

Powdery mildews are some of the most important plant pathogens, causing damage to crops such as beet, cereals, cucurbits, hops, ornamentals, soft fruit, tree crops, vegetables and vines. One of the most significant is perhaps *Erysiphe graminis* which causes considerable losses to European cereal crops. These pathogens are placed in the class Plectomycetes since they form an ascocarp (*cleistothecium*) which is closed and globose, not opening by an ostiole. They are biotrophic parasites usually forming a superficial mycelium from which haustoria are produced into the host epidermal cells. A few members of this group, such as *Leveillula taurica* which affects tomato and pepper in Mediterranean areas, form a mycelium within the host tissues. *Sphaerotheca pannosa* infects *Rosa* spp. and some other members of the Rosaceae wherever they are grown. As such it is an important and economically damaging pathogen of a major plant in the nursery-stock industry.

Symptoms

Symptoms typical of a powdery mildew are associated with *S. pannosa*. These begin as discrete star-shaped patches of mycelium on the lower leaf surface and spread over the youngest leaves leading to distortion, curling and premature leaf fall. At an early stage of infection, the stems become invaded, especially at the base of thorns where fungal growth persists as they mature. Flowers may be infected leading to a formation of small round patches of thin mycelium which cause petal spotting. The pathogen grows abundantly on pedicels, sepals and receptacles. If blooms are attacked prior to opening, then flower quality is reduced. Where infections persist and develop, eventually a dense white to grey or buff mycelial mass covers stems, thorns, leaves, calyces, flowers and hips and becomes especially abundant on receptacles. Severe infection reduces leaf growth and causes flower death, and may result in shoot death. Two types of *S. pannosa* mycelium may be distinguished: a thin sometimes transitory mycelium on young leaves and a more persistent coarser form consisting of thick-walled hyphae growing in tough felty patches known as 'pannose mycelium' from which the species name is derived.

Biology

The major reproductive phase of *S. pannosa* is asexual, by means of conidia. Mycelial formation on the host surface has been well documented for several powdery mildews. Germination of conidia of *S. pannosa* is stimulated by high humidity (RH 97–99 per cent) and temperatures at about 20°C. Subsequent development into a mycelium is retarded by free-water films on the leaves particularly if wetting takes place just after conidial deposition. A short primary germ tube which elongates and enlarges is produced from each conidium within 2–4 h after deposition.

By 6 h after deposition a club-shaped appresorial initial is produced which adheres the tube to its substratum: by 8–10 h a second germ tube forms usually from the opposite end of the conidium. By 16–20 h haustorial initials develop and by 20–24 h mature haustoria are found within the host epidermal cells. There follows rapid growth of second and third germ tubes; by 48 h conidiophore initials develop and by 72 h conidial chains are produced. Consequently, generation time from conidial deposition through mycelial formation to the production of further conidia is rapid, leading to fast spread of *S. pannosa* (Wheeler, B. E. J., 1978). Growth of this fungus on the host surface is related to both temperature and RH but temperature is the more important factor with optimal growth at 18–25°C and sporulation at 21–27.5°C (figure 2.9). Conidial production is encouraged by low RH with peak release of conidia in conditions of decreasing RH. In general the most favourable conditions

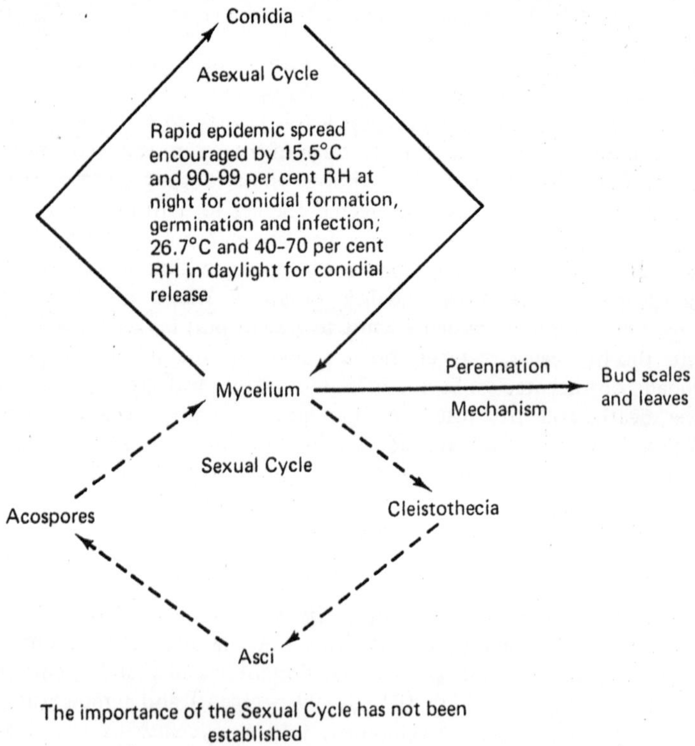

Figure 2.9 Infection cycle of *Sphaerotheca pannosa*—powdery mildew of rose and some other Rosaceae

for *S. pannosa* are 15.5°C and 90-99 per cent RH at night for conidial formation, germination and infection and 26.7°C and 40-70 per cent RH in daylight for release of conidia (Wheeler, B. E. J., 1973, 1975).

There are indications that, in leaves at least, rose tissue becomes more resistant to invasion with age; this may be associated with increased thickening of the leaf cuticle and outer epidermal wall in mature tissue. Infection may be related to growth periods of the host, being encouraged when there are large amounts of young easily penetrated tissue available. Pedicels are prime sites for growth by *S. pannosa* and act as inoculum sources for infection of other organs.

Cleistothecia form most abundantly in the pannose mycelium but formation is erratic and in some geographical areas is reputedly absent. They are most frequently found on rambler, climber and shrub types of rose and their value as perennation mechanisms has not been established. There are few reports of ascospore discharge from cleistothecia. It has been suggested that a more important source of overwintering and primary inoculum emanates from mycelium on bud scales and leaves.

Control

Resistance to this pathogen is continually sought by rose breeders. New cultivars often exhibit high levels of resistance in trials but after a few years in more general cultivation become heavily infected. Presumably this is due to an increased frequency of virulence factors in the pathogen population, which are compatible with resistance in a new cultivar, developing as it becomes cultivated on a wider scale with consequent erosion of resistance (section 5.6).

Control of *S. pannosa* by chemicals was advocated as early as 1860 when copper sulphate sprays were recommended. Since then control by this means has followed trends with other crops through materials such as sulphur and its derivatives, refined copper compounds (which are less phytotoxic), the dithiocarbamates to systemic and other materials. Widespread use has been made of the systemic fungicides benomyl and dodemorph for control of *S. pannosa*. In woody species, such as rose, they are not translocated to any extent, but their absorption into leaf tissue means that they are less affected by weathering and require fewer applications. Strains of *S. pannosa* tolerant to benzimidazole compounds have been reported. Major use is made of the non-systemic protectant dinocap and the more recently developed systemic compounds bupirimate and imazilil.

Pruning can be used to remove sources of overwintering infection particularly on hybrid tea (large-flowered bush rose) cultivars. Those rose types that are damaged by harsh pruning will benefit by removal of mummified flower heads where the receptacle bears pannose mycelium. It is necessary to remove all prunings from rose beds and preferably to burn them.

2.6.7 *Armillaria (Armillariella) mellea* (honey fungus)—a cause of dieback, root and butt rot of many tree and shrub spp.

A wide range of morphological form is found within the subdivision Basidiomycotina varying from edible and poisonous mushrooms and toadstools to important pathogens such as rusts and smuts. Throughout this group the mycelium consists of branched, septate hyphae often with perforated cross-walls (*septa*) similar to those found in the Ascomycotina. Vegetative fusion between separate hyphal strands (*anastomosis*) converts the mycelium of many Basidiomycotina into three-dimensional structures and characteristic clamp connections develop at the septa in many species. The most typical feature of this group is the sexual spore-producing structure (*basidium*) which usually carries four spores (*basidiospores*). In comparison to the ascus of the Ascomycotina where ascospores develop internally, the basidium produces basidiospores externally. Like the Ascomycotina, spore-producing basidia are massed together into a fertile hymenial layer (figure 2.10). Asexual reproduction in the Basidiomycotina occurs by several means such as budding, mycelial fragmentation and formation of asexual conidia; the latter are sometimes referred to as *oidea*. Characteristics of some classes within the Basidiomycotina are as follows.

Hymenomycetes

The hymenium in members of this class is exposed at maturity and borne on large easily recognisable fruiting bodies (*basidiocarps*) such as toadstools, mushrooms and fungal brackets which discharge basidiospores explosively. Examples include *Armillaria mellea* (honey fungus)—a serious pathogen of many dicotyledonous hosts, *Heterobasidion annosum*—the cause of heart rot or butt rot of conifers, *Polyporus betulinus*—the cause of heart rot of birch, *Chondrostereum purpureum*—the cause of silver leaf disease in rosaceous hosts, *Exobasidium* spp. which cause galling on many *Ericaceae* and *Thanetephorus cucumeris* the perfect stage of *Rhizoctonia solani* which is a notorious pathogen of a wide variety of horticultural crops causing root rots, cankers, damping-off, fruit decay and foliar diseases.

Gasteromycetes

Basidia are not exposed at maturity in members of the Gasteromycetes and the basidiospores are not discharged explosively. The basidiocarp is an enclosed structure, for example, puff balls, stinkhorns and bird's nest fungi. All members of this class are saprophytes of substrata such as soil, rotting wood, other vegetation and dung. Characteristically the basidiocarps have a distinct outer wall which may open naturally once the basidiospores mature or remain permanently closed, the basidiospores being liberated only after disintegration of this wall by external agencies.

Teliomycetes

Two important groups of plant pathogens compose the Teliomycetes: the smut fungi (Ustilaginales) and the rust fungi (Uredinales). In members of this class the hyphal septa are simple and lack pores; the basidium is a thick-walled resting spore (*teliospore*) within which nuclear fusion takes place and on germination forms a promycelium that gives rise to four or more basidiospores. Typically in the Uredinales only four basidiospores develop on the promycelium while in the Ustilaginales the number is large and indefinite. As plant parasites the Ustilaginales are confined to angiosperm hosts but the Uredinales parasitise pteridophytes, gymnosperms and angiosperms having in many cases complex life cycles which are completed on two botanically unrelated hosts. Examples of smut fungi (Ustilaginales) are *Ustilago maydis*—maize smut, *U. nuda*—loose smut of barley and *Tilletia caries*—the cause of bunt of cereals. Examples of rust fungi (Uredinales) are *Puccinia graminis*—the cause of black stem rust of cereals and grasses, *Cronartium ribicola*—the cause of blister rust of five-needled pines such as *Pinus strobus*, the white pine, and currants (*Ribes* spp.) (section 2.6.8), *Phragmidium mucronatum*—rose rust, *Uromyces fabae*—rust of *Vicia* beans and *U. dianthi*—rust of *Dianthus* and *Euphorbia* spp.

Armillaria (Armillariella) mellea is a root-invading pathogen and one of the commonest causes of death amongst trees and shrubs in Britain and virtually throughout the world. It may also cause wind-blowing and butt rot when infected plants are still alive. The specific name derives from the honey-coloured (melleus) toadstools that grow on and around infected trees; another name is 'bootlace' fungus because of the black strands (*rhizomorphs*) by which this pathogen spreads through soil from one host to another.

Symptoms

Within ornamental plantings of trees and shrubs, death due to *A. mellea* occurs in isolated individuals; rarely are whole woodland plantations affected at the same time. Greatest damage is done in parks and arboreta because of the high intrinsic and aesthetic value of each plant. Death may occur suddenly or proceed more slowly from a gradual dieback. Often infected hosts form fewer leaves than healthy ones although conifers may form an unusually heavy crop of cones when infected. Resinous, gummy exudates form at the base of infected plants. This pathogen may be identified from the creamy-white, skin-like mycelium which is produced beneath the host bark. Rhizomorphs are often difficult to find and recognise in soil. Very characteristic basidiocarps develop in early autumn, late September to early October, and aid identification. The cap (*pileus*) is 5–15 cm diameter, dome-shaped at first becoming flattened with a central depression varying in colour from olive-yellow to tawny-deep brown. The stem (*stipe*) is 7–15 cm long and 1 cm or more thick, sometimes bulbous at the base and of a similar colour to the cap. A ring (*annulus*) of whitish, thickened tissue is attached to the upper stem. Within the cap are whitish flesh-coloured gills (figure 2.10).

Butt rot of infected hosts can rarely be seen externally but when infected stems are cut the developing decay is seen as irregular pale-brown

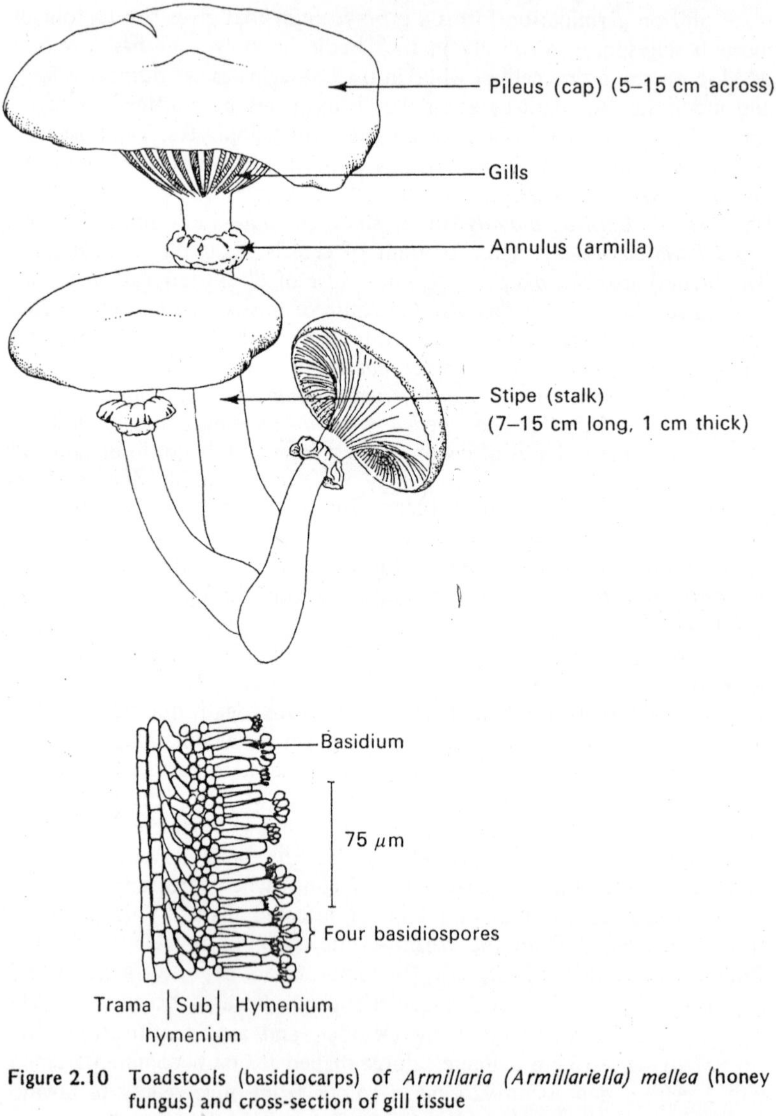

Figure 2.10 Toadstools (basidiocarps) of *Armillaria (Armillariella) mellea* (honey fungus) and cross-section of gill tissue

or inky-blue-grey patches of stain near the stem centre. Once the woody tissues have begun to break down, their centres are orange-brown and in the final stages of decay become wet, stringy brown and rotten with pale-yellow flecks. Characteristically, although decay is extensive, it is rarely found more than 50 cm above soil level.

Biology

This necrotrophic organism causes lethal damage by destroying the cambial layers in roots and lower stems of hosts. Thereafter it becomes saprophytic colonising and rotting woody tissues after death of a host and it may cause butt rot without invading living host tissue. Where colonisation is well advanced, masses of creamy-white mycelia are found beneath the bark. As host nutrient reserves are depleted the hyphae aggregate into rhizomorphs; at first these are yellowish but turn reddish-brown and finally black at maturity. They have a slightly elastic branched structure with a hard black exterior and white core. Rhizomorphs are capable of growing at least 1 m per year through soil to invade new hosts. Their distribution is related to soil moisture, being concentrated in the upper soil when this is moist but at deeper levels when the upper soil is dry. Rhizomorphs grow along gradients of increasing oxygen and decreasing carbon dioxide concentration. Soil disturbance stimulates their growth (Redfern, 1973) hence infection levels may increase as a result of tree-felling. Growth is retarded by high temperatures.

The basidiocarps (*toadstools*) are initiated from soil-borne rhizomorphs. Most cells in the mycelium, basidiocarp primordia, stipe and pileus (figure 2.11) are monokaryotic and thought to be diploid. Cells of the gill tissue are dikaryotic and thought to be haploid. Basidiocarp tissue is pseudoparenchymatous, consisting of aggregations of hyphae. Three layers of tissue may be distinguished in the gills: a loosely packed central layer (*trama*), a more densely packed *sub-hymenium* and an outer layer (*hymenium*) composed of basidia and basidiospores (figure 2.10). This fungus is thought to be homothallic in that mycelium derived from a single spore is capable of fruiting. Nuclear fusion and meiosis take place in the basidia. A single meiotic product enters each basidiospore and there divides mitotically. One daughter nucleus from this division remains in the spore; the other migrates back into the body of the basidium and degenerates (Ullrich and Anderson, 1978).

Transmission of *A. mellea* takes place by airborne basidiospores but the exact role that these play in dissemination is uncertain; they are incapable of penetrating directly into living tissue but can establish saprophytic colonies on dead plants, posts and fences. From these, pathogenic colonisation may spread by rhizomorphs into living tissue. Other modes of transmission are by natural grafting between healthy and infected host roots, via infected soil debris and with infected planting material. Optimal

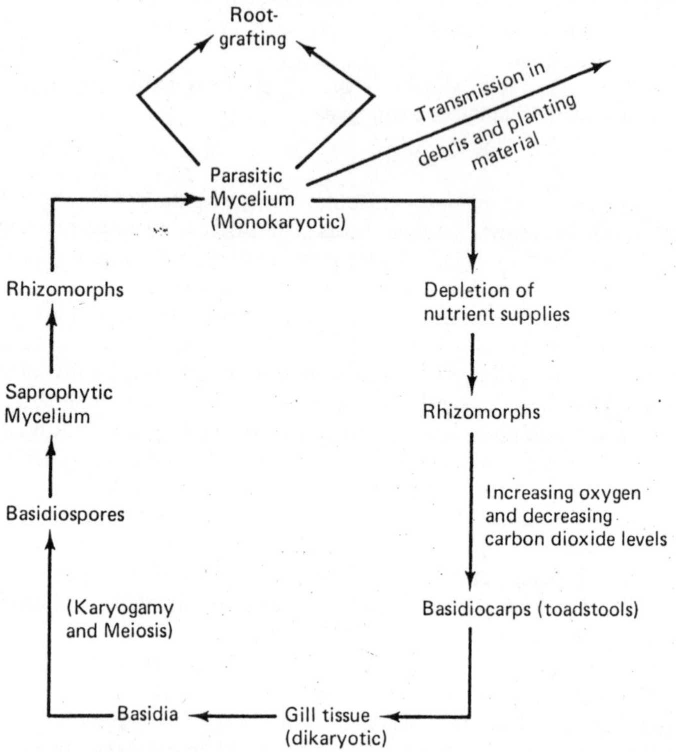

Figure 2.11 **Infection cycle of *Armillaria (Armillariella) mellea* (honey fungus)—a cause of dieback, root and butt rot of many tree and shrub spp.**

conditions for invasion by *A. mellea* are not known in detail; the severity of an attack depends on host susceptibility and vigour, the active mass of inoculum available and environmental conditions. Colonisation is favoured by high RH and low temperatures.

Control

A vast range of plant species is susceptible to *A. mellea* to some degree. In some, levels of resistance are known to increase with age. There is variation in the parasitic ability of strains of *A. mellea*. Stress from environmental factors will lead to death following invasion by *A. mellea*. Hosts that are notably susceptible and killed irrespective of age include: *Araucaria araucana* (monkey puzzle), *Betula* spp. (birch), *Cedrus* spp. (cedar), *Chamaecyparis* spp. and *Cupressus* spp. (cypress), *Juglans* spp., but not *J. hindsii*, (walnut), *Ligustrum* spp. (privet), *Malus* spp. (apple), *Picea omorika* (Serbian spruce), *Pinus* spp. (pine), *Salix* spp. (willow), *Sequoiadendron giganteum* (wellingtonia) and *Syringa* (lilac). Subjects that have shown some resistance and may be planted in infected areas with some degree of success are: *Abies* spp. (silver fir), *Ailanthus altissima* (tree of heaven), *Acer negundo* (box elder), *Arundinaria* and other genera (bamboo), *Buxus sempervirens* (box), *Cistus* spp. (rock rose), *Clematis* spp., *Cotinus coggygria* (Venetian sumach or smoke tree), *Crataegus* spp. (thorns), *Elaeagnus* spp., *Fagus sylvatica* (beech), *Fraxinus excelsior* (ash), *Hedera helix* (ivy), *Ilex aquifolium* (holly), *Juglans hindsii* (Californian black walnut), *Larix* spp. (larch), *Liquidamber styraciflua* (sweetgum), *Lonicera nitida*, *Mahonia* spp. (Oregon grape), *Quercus* spp. (oak), *Prunus laurocerasus* (cherry laurel), *Prunus spinosa* (blackthorn or sloe), *Pseudotsuga menziesii* (Douglas fir), *Rhus typhina* (stag's horn sumach), *Robinia pseudacacia* (locust tree or false acacia), *Sambucus niger* (elder), *Tamarix* spp. (tamarisk), *Taxus baccata* (yew) and *Tilia* spp. (lime). Butt rot symptoms are especially common in *Picea* spp. (spruce), *Tsuga canadensis* (western red cedar) and *Chamaecyparis lawsoniana* (Lawson's cypress).

Where infected ornamental plants are removed, apparently healthy ones in close proximity should also be taken out to deprive the pathogen of food bases. Infected stumps should be removed but if this is impossible they should be chipped *in situ* by machine and the chips removed and burnt. Stump removal is particularly worth while in nurseries, orchards and gardens.

Sterilisation with carbon disulphide gas is used to clean up infected land in Californian fruit orchards; this practice is uncertain and not suited to Britain but other gaseous soil sterilants, such as methyl bromide, could be used. Biological control may be possible by the encouragement of competitive fungi that replace *A. mellea*. These may be stimulated by treating stumps with ammonium sulphamate and inoculating with competitors such as *Coriolus versicolor* or *Phlebia merismoides* (Risbeth, 1976).

2.6.8 *Cronartium ribicola*—blister rust of *Pinus* and *Ribes* spp.

Rust fungi (Uredinales) are examples of biotrophic pathogens forming an intercellular mycelium from which haustoria develop into neighbouring cells. Unlike many members of the Basidiomycotina the rusts do not produce basidiocarps but instead sporulating lesions initiate within the host and erupt through the epidermis releasing several types of spore. Up to five spore types may be produced by a rust fungus in its life cycle. The specialised spore bodies are denoted by the numerals 0 to IV.

Spermogonia (0)

Infection by basidiospores or sporidia (stage IV) results in the formation of spermogonia. They develop from a uninucleate mycelium and vary in shape and size from flask-like through conical to flat and sprawling. When mature, spermogonia release *spermatia* in a sweet, sticky exudate which is attractive to insects. These transfer the spermatia to receptive hyphae within spermogonia of opposite mating type (most rusts are heterothallic). Spermatial nuclei move into the receptive hyphae forming dikaryotic cells at the bases of the spermogonia.

Aecia (I)

This stage results from dikaryotic cells formed within the fertilised spermogonia; aecia develop on the side of host organs away from the spermogonia. Usually they are seen as blisters on the undersurfaces of leaves and contain chains of *aeciospores*. The latter are dikaryotic propagules which transfer the fungus from one host species to another.

Uredia (II)

Infection from aeciospores normally takes place through stomata to form a dikaryotic mycelium that produces uredial pustules containing *uredospores*. These are a repeating stage within the life cycle, often producing many further generations of uredospores. Typically these are thick-walled and spiny, viable for several weeks, and serving to propagate the fungus on its host and permitting spread to neighbouring plants or long-range dissemination.

Telia (III)

In many rusts this stage develops within existing uredia. Owing to their perfect (sexual) state and wide morphological range, telia and their resultant spores are used to classify the rusts. The thick-walled *teliospores* (or *teleutospores*) are at first dikaryotic but by karyogamy each becomes diploid and uninucleate. They have a perennating and overwintering function.

Basidia (IV)

At germination each teliospore produces a tube into which the nucleus migrates; by meiosis four haploid nuclei form and four uninucleate cells are cut off by septal development. From each cell a *basidiospore* forms containing a nucleus. The basidiospores are discharged into air currents and transfer the rust to another host species. Basidiospore germination may be either direct by means of a germ tube or by production of a further crop of spores (*sporidia*). In either case host invasion results in a mycelium that produces spermogonia (stage 0) (Petersen, 1974).

Most rusts possess the perfect, telial, stage and produce spermogonia. The other spore stages, which may be viewed as imperfect states, may be omitted. On these mycological criteria, rusts are classed as one of the following.

1. *Macrocyclic*—where all five reproductive states are present.
2. *Demicyclic*—where the uredinal state is absent.
3. *Microcyclic*—where teliospores are the only binucleate spore produced.

In relation to pathogenicity on host plants, rusts may be classed as one of the following.

1. *Heteroecious*—parasitising two botanically distinct hosts; stages 0 (spermogonia) and I (aecia) are found on one host and stages II (uredia), III (telia) and IV (basidia) on another. The host on which telia develop is the *primary* host and that bearing spermogonia and aecia is the *alternate* host.
2. *Autoecious*—in which the entire life cycle is completed on a single host species.

Cronartium ribicola is a macrocyclic heteroecious rust producing spermogonia and aecia on five-needled pines (*Pinus* spp.), especially *P. strobus* the white pine, and its other spore bodies on *Ribes* spp., particularly blackcurrant (*R. nigrum*) and gooseberry (*R. grossularia*) (figure 2.12). This parasite existed in a balanced ecology in Europe, north-east Asia and Siberia prior to the nineteenth century when *P. strobus* was introduced from North America to Europe. This provided a highly susceptible host and led to epidemic rust development in European plantations of *P. strobus*. Around 1900 some infected seedlings of *P. strobus* were exported from Europe to the USA as cheap planting material and in about 1920 similar exports were made to Canada. Subsequently forests of *P. strobus* and *P. monticola* in North America became seriously infected with *C. ribicola*.

Symptoms

Pine needles are first infected in late summer or early autumn by basidiospores released from the primary *Ribes* host. Subsequently yellow-to-brown lesions develop on infected needles. The bark of twigs bearing diseased needles turns yellowish and becomes swollen during the following summer. Spermogonia first appear in the bark of swollen twigs during the summer following infection or in the early spring in the subsequent year. These develop into amber-coloured blisters which rupture releasing spermatia in a sugary liquid. Aecia form in the spring in old spermogonial scars while further spermogonia develop at the periphery of swollen parts of stems. The aecia cause white elongated blisters (hence the name white blister rust) to develop which rupture in spring exposing masses of orange-yellow aeciospores. General symptoms on pine stems consist of a central area of dead bark surrounded by zones of aecia; spermogonia develop only between the aecia and in the advancing margins of infection

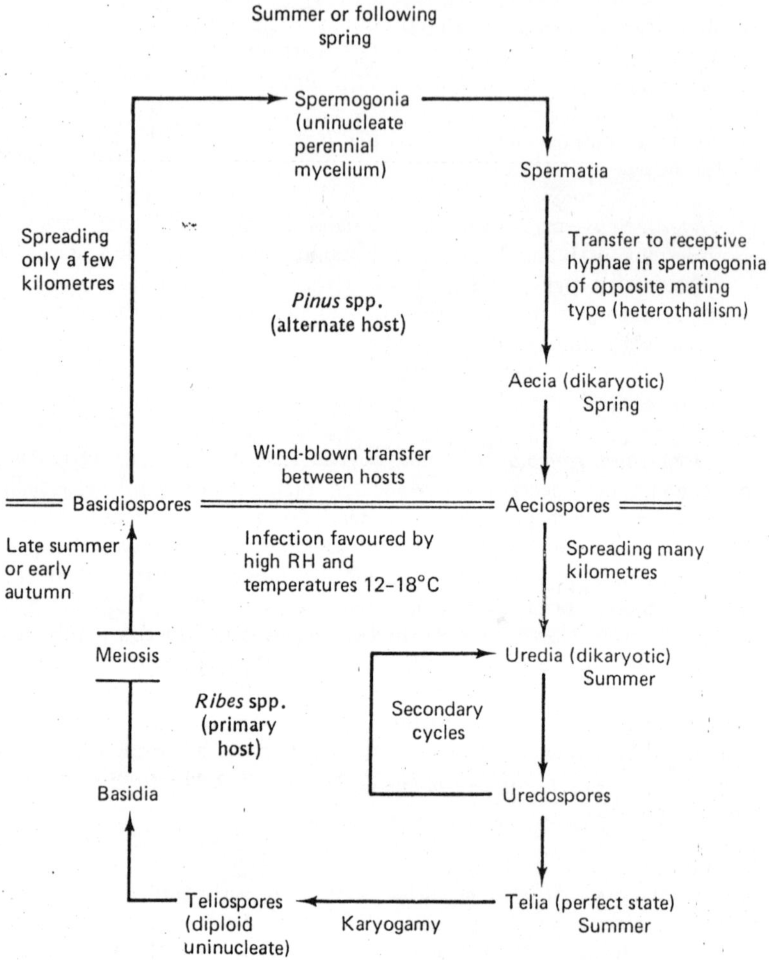

Figure 2.12 Infection cycle of *Cronartium ribicola*—the cause of blister rust of *Pinus* and *Ribes* spp.

(Miller *et al.*, 1959). In this pathogen spermogonia and aecia develop together, not separately as is more common with rusts.

Aeciospores blown by wind from infected pines initiate infections on currant and gooseberry. Uredial lesions form mainly on the undersurface of *Ribes* leaves, but young twigs, buds and floral organs are occasionally invaded. Telia develop within the uredia. Chlorotic yellow-to-brown flecks can be found on the upper surfaces of leaves above these fruiting bodies.

Repeated annual infections cause premature defoliation of *Ribes* bushes, weakening and eventually killing them. The fungus is not thought to perennate in a mycelial form on *Ribes* hosts over winter.

Biology

Since infection in *Pinus* spp. spreads by mycelial growth from the needles to stems, species such as *P. monticola* which retain their needles for several years are severely injured by *C. ribicola*. The number of stomata per unit area of leaf in different *Pinus* spp. is also thought to be a factor in their relative susceptibility. Growing in from infected leaves towards the stem, the mycelium enters the intercellular spaces of the cortex and phloem becoming well established between these two tissues and wedging them apart so that the phloem is crushed against the wood. Accordingly, cambial division in infected zones is less active than in healthy stems but the secondary wood is normal except that thinner increments of tracheids are added to the annual rings. Once mycelium accumulates between the cortex and phloem, the bark bursts and a canker develops. This leads to exposure and drying out of delicate tissues followed by splitting and copious exudation of *Pinus* resin around the canker. Extension of cankers around twigs and branches leads to girdling with periodic swelling and drying of the exudate. Impregnation of the cortex and phloem near the canker, with resin, stops translocation leading to hypertrophy around the infected area. On trees less than five years old cankers may appear within six months of infection whereas in older trees they may take two years to become visible. Where girdling progresses down a tree whole branches or trees may be killed. Stubs of cankered branches can remain alive and harbour infection for up to five years. The mycelium that produces spermogonia and aecia in *Pinus* spp. is uninucleate and perennial; that arising from aeciospore infection in *Ribes* spp. is binucleate. The uredia are minute domed structures whereas the telia contain elongated horn-like columns of teliospores.

Transfer of *C. ribicola* from *Pinus* to *Ribes*, and vice versa, is accomplished by wind-blown aeciospores and basidiospores respectively. Conditions that favour germination of these propagules are periods of dew, heavy fog or cloudy weather and temperatures of 12–18°C. At least 48 h of high RH are needed for basidiospore production and subsequent infection of *Pinus* spp. Aeciospores released from *Pinus* are capable of being transported by wind for many kilometres but spread by basidiospores from *Ribes* is seldom effective over distances of more than 1.5 km.

Control

The blackcurrant (*R. nigrum*) and flowering currant (*R. sanguineum*) are very susceptible to *C. ribicola* whereas redcurrant (*R. sativum*) and gooseberry (*R. grossularia*) are slightly resistant. Among the five-needled pines

P. strobus, *P. monticola*, *P. flexilis* and *P. lambertiana* are highly susceptible. *Pinus excelsa* and *P. cembra* are less susceptible and *P. peuce* is moderately resistant. Some work has been done in Canada to develop resistant currants and pines. Selections of *P. monticola* and *P. strobus* have been propagated together with hybrids using *P. peuce* and *P. excelsa*. Legislation in North America prevents importation of *Ribes* and *Pinus* material and vast amounts of money have been spent to eradicate currants and gooseberries from pine forest areas. For instance, by 1942 over 11 million hectares in the Eastern States had been cleared of wild and cultivated *Ribes* spp. (Fulling, 1943). In New York State, however, *Ribes* crops were more valuable than pines and accordingly the forest trees were removed. More recent eradication programmes have used aerial sprays of hormone herbicides such as 2,4-D and 2,4,5-T to eradicate *Ribes* spp. In Britain, garden currant bushes are the main sources of infection and are virtually impossible to eliminate. This has resulted in the loss of *P. strobus* and other five-needled pines to forestry, preventing the use of these valuable trees. Pruning trees of ornamental significance to remove infected areas can be successful especially if pruning scars are treated with cycloheximide which apparently kills *C. ribicola* mycelium and allows the wounds to heal. Chemical control with Bordeaux mixture applied at bud burst has been advocated for *Ribes* spp. but more effective control may be achieved with sprays of mancozeb, manganese–zinc dithiocarbamate or zineb.

2.6.9 *Verticillium albo-atrum* and *V. dahliae*—causes of vascular wilt diseases in a range of crops

The subdivision Deuteromycotina is an artificial grouping of those fungi which reproduce only by asexual means. They may be considered as conidial or imperfect stages of Ascomycotina or, less frequently, Basidiomycotina whose sexual stages either have not been discovered or do not exist. Conidia within this group have a tremendous range of morphological form; they may be, for example, spherical, ovoid, elongated, cylindrical, thread-like, spirally curved or star-shaped. Conidia may have one or more cells with either transverse septa or both transverse and longitudinal septa; they may be colourless and transparent (hyaline) or coloured, and develop singly or in chains. An important distinguishing feature related to dispersal is whether conidia are dry or sticky. Dry-spored conidia are generally wind dispersed; where the conidia have slimy coats they cohere, forming aggregations or sometimes tendril-like ribbons and the spores may be dispersed by insects or by rain-splash.

Conidia may form on single stalks (*conidiophores*) or the reproductive hyphae may be grouped together to form specialised asexual structures such as the following.

Synnemata (Coremia)
These form from a group of conidiophores often united at the base and part of the way up with conidia produced along its length.

Sporodochia
In these structures conidiophores arise from a central cushion-shaped mass of compacted mycelium.

Pycnidia
Globular or flask-shaped pseudoparenchymatous structures lined on the interior by conidiophores are termed pycnidia; they superficially resemble the perithecia of Ascomycotina. Pycnidia are, however, asexual structures rather than sexual, producing conidia as opposed to asci containing ascospores (figure 2.13). They may be completely closed or have an opening or ostiole.

Acervuli
These consist of flat or saucer-shaped structures formed of short conidiophores developing from a stromatic mass of hyphae. They form subepidermally or subcutically in the host plant tissue and erupt at maturity above the host surface releasing conidia (figure 2.13).

Three classes are identified within the Deuteromycotina: *blastomycetes* which are imperfect yeast-like organisms where a mycelium is absent or not well developed, for example, *Itersonilia pastinacae* (a cause of parsnip canker) and *I. perplexans* (chrysanthemum petal blight) and the *coelomycetes* and *hyphomycetes* which form a septate mycelium. Orders within the latter two classes are differentiated on the form of the asexual spore body as follows.

Coelomycetes
1. *Sphaeropsidales*—where the conidia form within a pycnidium, for example, *Septoria apiicola* (celery leaf spot) and *Phoma exigua* f. sp. *foveata* (potato gangrene).
2. *Melanconiales*—where conidia form within an acervulus, for example, *Colletotrichum lindemuthianum*, the cause of anthracnose in *Phaseolus* bean.

Hyphomycetes
Moniliales—where conidia are produced in sporodochia, synnemata or as discrete conidiophores. Those forming the latter are more important as plant pathogens; examples are *Alternaria* spp. causing leaf spots on brassicas, *Fusarium solani* f. sp. *phaseoli* and *F. solani* f. sp. *fabae* (root rot of beans) and *Verticillium albo-atrum* and *V. dahliae*.

Some fungi form only grey sterile mycelia producing neither asexual

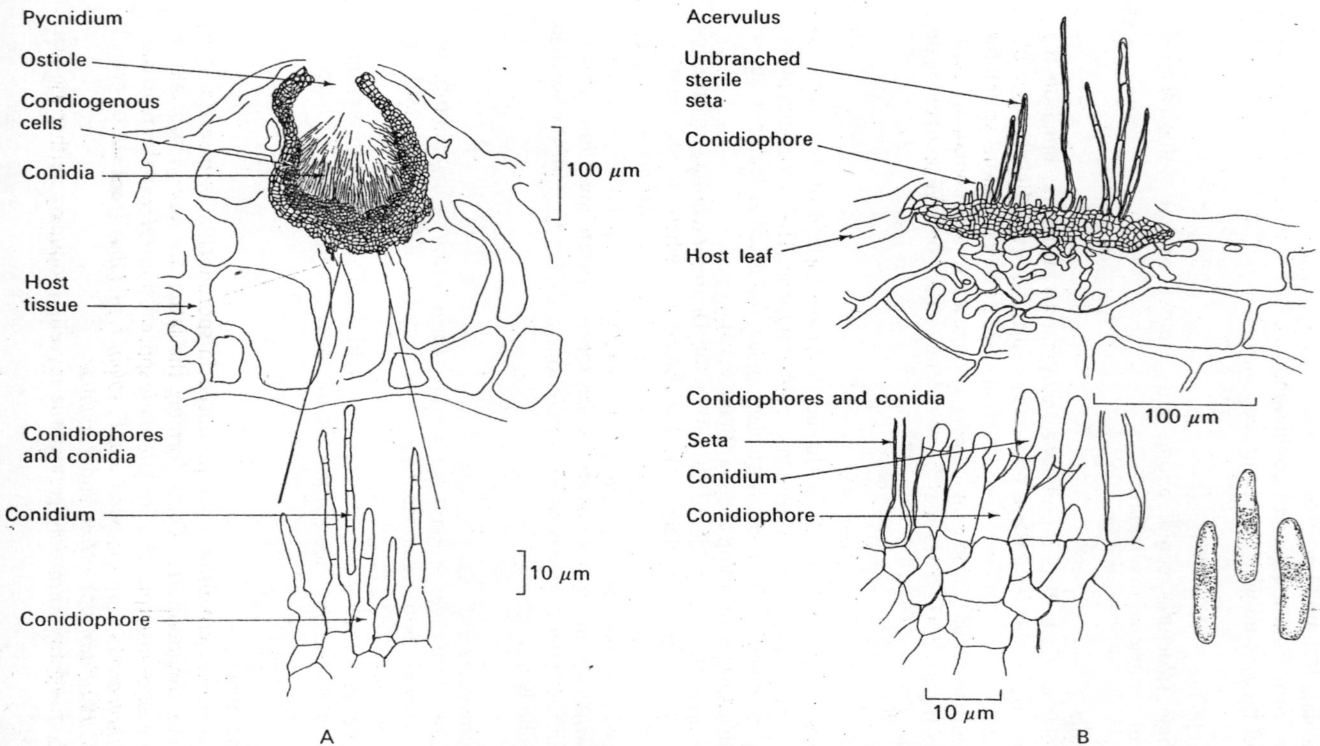

Figure 2.13 Examples of some asexual structures within the Deuteromycotina. (A) Pycnidium: from *Septoria lycopersici* (cause of tomato leaf spot) after Sutton and Waterston (1966). (B) Acervulus: from *Colletotrichum coccodes* (cause of black dot, anthracnose

nor sexual spores but reproducing by random hyphal fragmentation in the wild. These organisms are referred to as *Mycelia sterilia.* When cultured artificially, spore formation may be induced and a more satisfactory identification obtained. An example is *Pyrenochaeta lycopersici*, the cause of brown or corky root rot of tomato, which by induction of pycnidial and conidial formation in culture has been identified at least as a member of the Sphaeropsidales.

Species of *Verticillium*, particularly *V. albo-atrum* and *V. dahliae*, have a world-wide distribution in both temperate and tropical regions, causing very significant damage to many crops. These two species are considered together because it may be disputed as to whether or not they are separate organisms. Hosts include trees, shrubs and perennial and herbaceous types of horticultural crops including tomato, hop, pepper, potato, strawberry, raspberry, mint, olive and water melon. *Verticillium albo-atrum* is of greatest significance on hops and solanaceous crops while *V. dahliae* is of greatest significance on strawberry and stone fruits.

Symptoms

Both *Verticillium* species are soil-borne, invading the host through root tissues, penetrating the cortex and entering the vascular system; hence the general term vascular wilt. The fungus only re-emerges from the vascular system when the host becomes moribund. Once established within the root xylem vessels, colonisation ensues by mycelial growth and by transport of conidia in xylem fluids. The resultant effects are to impede passage of water within the host, leading to wilting. At first, wilting is reversible and the plant regains turgidity at night but gradually wilting becomes irreversible. This is accompanied by foliar yellowing (*chlorosis*) which is first seen in the lower leaves but slowly becomes generalised and accompanied by death of the laminae (*necrosis*). Leaves show down-bending symptoms (*epinasty*) which are attributed to hormonal disturbances within the host. The leaflets turn a dull green and curl inwards at the margins. Chlorotic areas on the leaves become crisp and eventually wither and desiccate. On hop leaves in particular 'tiger-striping' symptoms develop which consist of clear yellow areas separating marginal and interveinal zones of dark necrotic tissue from persistently green areas along and adjacent to the main veins, giving a yellow and black appearance. Symptom expression in hop is best described as progressive desiccation rather than wilting. In most hosts vascular discolouration is seen within roots, stems and petioles; adventitious roots develop profusely from the stem base especially in herbaceous types. Symptoms other than vascular browning are found within the vessels when examined microscopically; deposits of gums and gels form in vessel lumens while tyloses are found expanding into vessels. Tyloses are balloon-like projections that extend from xylem parenchyma cells through the vessel-wall pits into the vessel cavity thereby possibly blocking the passage of xylem fluids and movement of the fungus. There is considerable disruption of host hormone levels with increased levels of indole auxins and release of ethylene within diseased plants. Increased levels of nitrogen, especially amino compounds (Dixon and Pegg, 1972) and enzyme activity (Cooper and Wood, 1973) occur in infected plants. Toxins have been implicated as agents in the eventual death of some hosts. Infected plants are stunted with reduced leaf area and dry weight when compared with healthy ones.

Biology

Verticillium spp. reproduce asexually producing conidia in verticillate whorls about the length of the conidiophores. Both *V. albo-atrum* and *V. dahliae* produce resting hyphae as a means of perennation. Those formed by *V. albo-atrum* tend to be dark-brown or black beaded swellings of individual hyphae without forming microsclerotia, by repeated budding from single hyphae (Isaac, 1967). Transmission of conidia is in air currents and by water-splash as mycelium on infected plant debris. Some movement may occur via fine particles of infected host material accompanying or adhering to seeds. Transfer of these pathogens also takes place by infected soil carried on implements and in drainage water (figure 2.14). The two species, *V. albo-atrum* and *V. dahliae* differ slightly in optimal temperature requirements for growth ; *V. albo-atrum* grows at 20–25°C whereas *V. dahliae* thrives at 25–28°C. In practice this means

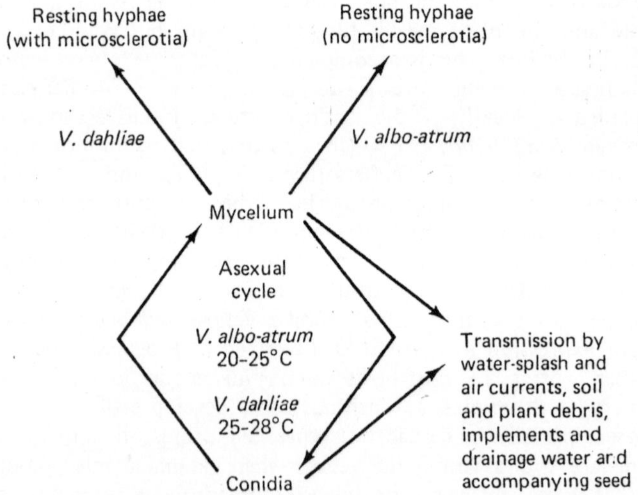

Figure 2.14 Infection cycle of *Verticillium albo-atrum* and *V. dahliae*—causes of vascular wilt diseases in a range of crops

that *V. albo-atrum* predominates in cool temperate climates such as northern Europe but *V. dahliae* predominates in warm temperate climates such as southern Europe, Mediterranean zones and the west coast of the USA. Symptoms become most apparent at low humidities when the host is under natural water stress.

Both pathogens exhibit necrotrophic and biotrophic features (table 1.1). Host range is wide and isolates may be transferred to different botanical species. Isolates from Brussels sprout, mint and pepper, however, tend to be more host specific. It is doubtful if these organisms can exist saprophytically for any length of time, although abundant sporulation is produced on moribund host tissues. Rate of death varies widely with different hosts but in distinction to other vascular pathogens, such as *Fusarium* spp., *Verticillium* spp. kill their hosts only slowly. An unresolved question is the means by which devastating symptoms develop in live host tissues as a result of a pathogen that exists in the inert xylem vessels.

Control

Resistance to *Verticillium* in tomato is of two types: a multigenic complexly inherited form from Manx Marvel and Bides' Recruit and a single dominant gene, Ve, obtained originally in the USA from a wild accession named Peru Wild. The latter gives resistance to *V. albo-atrum* and race 1 of *V. dahliae*. Many currently used commercial cultivars contain Ve gene but their resistance can be modified by temperature, soil pH and nutritional status. Resistance to *V. dahliae* race 2 is of a polygenic type obtained from Heinz 1350. Resistance for potatoes has also been sought from wild material but the pattern of inheritance is not clear. Some lines with high yield and wilt tolerance, that is the ability to endure infection, have been bred. Resistance is present in 13 *Solanum* spp. *Verticillium*-resistant strains of peppermint (*Mentha piperita*) have been developed by irradiation mutation breeding and have also been identified in *M. niliaca* and a *M. piperita* x *M. crispa* hybrid. Efforts are being made to develop strawberry cultivars with resistance to *V. dahliae*. No truly resistant cultivars have been produced but tolerance to wilt symptoms has been identified in Redgauntlet and Talisman from the Auchincruive breeding programme.

Chemical control in field crops aims to reduce the quantity of inoculum in post-harvest plant debris, to lower the soil inoculum by fumigation and, in the growing crop, to limit fungal growth by systemically fungicidal compounds. Surface-flaming of *V. dahliae*-infected mint debris and potato haulm is reported from the USA. These techniques are combined with fumigation using materials such as chloropicrin, dichloropropene and methyl bromide. In Germany, land for strawberry production is sterilised by injection with mixtures of methyl iso thiocyanate, dichloropropene and chloropicrin. Systemic chemicals active against *Verticillium* include the benzimidazole types such as benomyl, thiabendazole and thiophanate-methyl. These need to be applied in large volumes of water as soil drenches and are especially appropriate for glasshouse crops such as tomato and chrysanthemum but may be used for field crops like strawberry with reduced water volumes.

Control by husbandry methods includes use of herbicides to prevent the growth of alternative weed hosts; this has been particularly effective in hops, where reduction in the rates of nitrogenous fertilisers also dramatically lowered the incidence of wilt disease. Rotation of susceptible crops with cereals such as wheat and barley lowers the soil-borne inoculum level. Under glass, use of sterilisation by steam, steam and air mixtures or chemicals is effective but expensive and infections are largely circumvented by crop production in artificial media such as peat bolsters, rock wool or nutrient film.

REFERENCES

AINSWORTH, G. C., SPARROW, F. K. and SUSSMAN, A. S. (eds) (1973). *The Fungi—Advanced Treatise*, vols IVA and IVB, Academic Press, New York

AIST, J. R. and WILLIAMS, P. H. (1971). The cytology and kinetics of cabbage root hair penetration by *Plasmodiophora brassicae*. *Canadian Journal of Botany*, 49, 2023–34

ALEXOPOULOS, C. J. and MIMS, C. W. (1979). *Introductory Mycology*, Wiley, New York

BEAUMONT, A. and STANILAND, L. N. (1933). Potato blight and the weather. *Annual Report Seale Hayne Agricultural College for 1932*, 26–32

BECKETT, A., HEATH, I. B. and McLAUGHLIN, D. J. (1974). *An Atlas of Fungal Ultrastructure*, Longman, London

BURCKHARDT, H. and FREITAG, E. (1969). Die *Phytophthora*-Negativ-Vorhersage. *Der Kartoffelbau*, 20, 176–80

BUSHNELL, W. R. and GAY, J. (1978). Accumulations of solutes in relation to the structure and function of haustoria in powdery mildews. 183–236, in *The Powdery Mildews* (ed. D. M. Spencer), Academic Press, London

CARTER, M. V. (1963). *Mycosphaerella pinodes*. II. The phenology of ascospore release. *Australian Journal Biological Sciences*, 16, 800–17

COOPER, R. M. and WOOD, R. K. S. (1973). Induction of synthesis of extracellular cell-wall degrading enzymes in vascular wilt fungi. *Nature, London*, 246, 309–11

CRUTE, I. R. and DIXON, G. R. (1981). Downy mildew diseases caused by the genus *Bremia* Regel. 421–60, in *The Downy Mildews* (ed. D. M. Spencer), Academic Press, London

DICKINSON, C. H. and LUCAS, J. A. (1977). *Plant Pathology and Plant Pathogens*, Blackwell, Oxford

DIXON, G. R. and PEGG, G. F. (1972). Changes in amino acid content of tomato xylem sap following infection with strains of *Verticillium albo-atrum*. *Annals of Botany* n.s., 36, 147–54

DIXON, G. R. and WRIGHT, I. R. (1978). Frequency and geographical distribution of specific virulence factors in *Bremia lactucae* populations in England from 1973 to 1975. *Annals of Applied Biology*, 88, 287–94

FULLING, E. H. (1943). Plant life and the law of man. IV. Barberry, currant and gooseberry and cedar control. *Botanical Review*, 9, 484–592

HAWKER, L. E. (1967). *Fungi*, Hutchinson, London

HIRST, J. M., STOREY, I. F., WARD, W. C. and WILCOX, H. J. (1955). The origin of apple scab epidemics in the Wisbech area in 1953 and 1954. *Plant Pathology*, 4, 91-6

HORSFALL, J. G. and COWLING, E. B. (1978). *Plant Disease–An Advanced Treatise*, vol. III. *How Plants Suffer from Disease*, Academic Press, New York

HORSFALL, J. G. and DIMOND, A. E. (1959). *Plant Pathology–An Advanced Treatise*, vol. I. *The Diseased Plant*, Academic Press, New York

HUSAIN, A. and KELMAN, A. (1959). Tissue is disintegrated 144-88. *Plant Pathology–An Advanced Treatise* (eds J. G. Horsfall and A. E. Dimond), vol I. *The Diseased Plant*, Academic Press, New York and London

INGOLD, C. T. (1971). *Fungal Spores–Their Liberation and Dispersal*, Clarendon Press, Oxford

ISAAC, I. (1967). Specialisation in *Verticillium*. *Annual Review Phytopathology*, 5, 201-22

JARVIS, W. R. (1980). Epidemiology. 219-50, in *The Biology of Botrytis* (eds J. R. Coley-Smith, K. Verhoeff and W. R. Jarvis), Academic Press, London

JARVIS, W. R. and BORECKA, H. (1968). The susceptibility of strawberry flowers to infection by *Botrytis cinerea*. Pers. Ex Fr. *Horticultural Research*, 8, 147-54

JEGER, M. J., SWAIT, A. A. J. and BUTT, D. J. (1982). Overwintering of *Venturia inaequalis*, the causal agent of apple scab, on different cultivars. *Annals Applied Biology*, 100, 91-8

JOHNSON, A. G., CRUTE, I. R. and GORDON, P. L. (1977). The genetics of race specific resistance in lettuce (*Lactuca sativa*) to downy mildew (*Bremia lactucae*). *Annals Applied Biology*, 86, 87-103

KENDRICK, B. (ed) (1971). *Taxonomy of Imperfect Fungi*, University Press, Toronto

MALCOLMSON, J. F. (1969). Races of *Phytophthora infestans* occurring in Great Britain. *Transactions British Mycological Society*, 53, 417-23

MICHELMORE, R. W. and INGRAM, D. S. (1980). Heterothallism in *Bremia lactucae*. *Transactions British Mycological Society*, 75, 47-56

MILLER, D. R., KIMMEY, J. W. and FOWLER, M. E. (1959). *White pine blister rust*. United States Department of Agriculture and Forestry Pest Leaflet 36

MILLS, W. D. and LA PLANTE, A. A. (1954). *Disease and insects in the Orchards*. Extension Bulletin Cornell Agricultural Experimental Station No. 711, pp. 21-7

MORDUE, J. E. M. (1967). *Colletotrichum coccodes*. *Commonwealth Mycological Institute Descriptions of Pathogenic Fungi and Bacteria* No. 131. Commonwealth Mycological Institute, Kew

PETERSEN, R. H. (1974). The rust fungus life cycle. *Botanical Review*, 40, 453-513

REDFERN, D. B. (1973). Growth and behaviour of *Armillaria mellea* rhizomorphs in soil. *Transactions British Mycological Society*, 61, 569-81

RISBETH, J. (1976). Chemical treatment and inoculation of hardwood stumps for control of *Armillaria mellea*. *Annals Applied Biology*, 82, 57-70

SAVAGE, E. J., CLAYTON, C. W., HUNTER, J. H., BRENNEMAN, J. A., LAVIOLA, C. and GALLEGLY, M. E. (1968). Homothallism, heterothallism and interspecific hybridisation in the genus *Phytophthora*. *Phytopathology*, 58, 1004-21

SIVANESAN, A. and WALLER, J. M. (1974). *Venturia inaequalis*. *Commonwealth Mycological Institute Descriptions of Pathogenic Fungi and Bacteria* No. 401. Commonwealth Mycological Institute, Kew

SUTTON, B. C. and WATERSTON, J. M. (1966). *Septoria lycopersici*. *Commonwealth Mycological Institute Descriptions of Pathogenic Fungi and Bacteria* No. 89. Commonwealth Mycological Institute, Kew

TALBOT, P. H. B. (1971). *Principles of Fungal Taxonomy*, Macmillan, London

TOMMERUP, I. C. and INGRAM, D. S. (1971). The life cycle of *Plasmodiophora*

brassicae Woron. in Brassica tissue cultures and in intact roots. *New Phytologist*, 70, 327-32
TOWNSEND, B. (1957). Nutritional factors influencing the production of sclerotia by certain fungi. *Annals of Botany*, n.s., 21, 153-66
TOWNSEND, B. and WILLETTS, H. J. (1954). The development of sclerotia of certain fungi. *Transactions British Mycological Society*, 37, 213-21
ULLRICH, R. C. and ANDERSON, J. B. (1978). Sex and diploidy in *Armillaria mellea. Experimental Mycology*, 2, 119-29
WEBSTER, J. (1980). *Introduction to Fungi*, Cambridge University Press, London
WHEELER, B. E. J. (1973). Research on rose powdery mildew at Imperial College. *Journal Royal Horticultural Society*, 98, 225-30
WHEELER, B. E. J. (1975). Rose powdery mildew: field trials with systemic fungicides 1970-2, *Annals Applied Biology*, 79, 177-88
WHEELER, B. E. J. (1978). Powdery mildews of ornamentals. 411-46, in *The Powdery Mildews* (ed. D. M. Spencer), Academic Press, London
WHEELER, H. (1975). *Plant Pathogenesis*, Springer-Verlag, Berlin
WILLETTS, H. J. and WONG, A. L. (1971). Ontogenetic diversity of sclerotia of *Sclerotinia sclerotiorum* and related species. *Transactions British Mycological Society*, 57, 515-24
WOOD, R. K. S. (1967). *Physiological Plant Pathology*, Blackwell, Oxford

FURTHER READING

Commonwealth Mycological Institute Descriptions of Pathogenic Fungi and Bacteria, published by Commonwealth Mycological Institute, Kew. Numbers: *Armillariella mellea* (Honey Fungus) **321**; *Bremia lactucae* (Lettuce Downy Mildew) **682**; *Cronartium ribicola* (White Pine Blister Rust) **283**; *Plasmodiophora brassicae* (Clubroot) **621**; *Sclerotinia fuckeliana* (Grey Mould) **431**; *Sphaerotheca pannosa* (Rose Powdery Mildew) **189**; *Venturia inaequalis* (Apple Scab) **401**; *Verticillium albo-atrum* and *V. dahliae* (Wilt) **255, 256**

ANON (1968). *Plant Pathologists' Pocketbook*. Commonwealth Mycological Institute, Kew
DIXON, G. R. (1981). *Vegetable Crop Diseases*, Macmillan, London
FORSBERG, J. L. (1976). *Diseases of Ornamental Plants*, University of Illinois Press, Urbana
GREIG, B. J. W. and STROUTS, R. G. (1977). *Honey Fungus*, leaflet no. 2. Department of the Environment/Forestry Commission. Her Majesty's Stationery Office, London
HORSFALL, J. G. and COWLING, E. B. (eds). *Plant Disease, An Advanced Treatise.* Vol. I. *How Disease is Managed* (1977). Vol. II. *How Disease Develops in Populations* (1978). Vol. III. *How Plants Suffer from Disease* (1978). Vol. IV. *How Pathogens Induce Disease* (1979). Vol. V. *How Plants Defend Themselves* (1980). Academic Press, New York
PEACE, T. R. (1962). *Pathology of Trees and Shrubs*, Oxford University Press, London
ROBERTS, D. A. and BOTHROYD, C. W. (1972). *Fundamentals of Plant Pathology*, Freeman, San Francisco
TATTAR, T. A. (1978). *Diseases of Shade Trees*, Academic Press, New York
WALKER, J. C. (1952). *Diseases of Vegetable Crops*, McGraw-Hill, New York
WESTERN, J. H. (ed.) (1971). *Diseases of Crop Plants*, Macmillan, London

3 VIRUSES

3.1 THE NATURE OF VIRUSES

Man has for centuries recognised the symptoms caused by viruses without realising their cause. The garish colour breaks in tulip flowers so highly prized in seventeenth-century Europe resulted from virus infection spread during vegetative propagation. It was not until the late nineteenth century, however, that evidence was produced that such symptoms were caused by pathogens. It was then shown that mosaic disease to tobacco could be transmitted to healthy plants by inoculation with sap from an infected plant. Proof that such symptoms were the result of particles smaller than those of bacteria was obtained when sap from mosaic-infected plants was passed through a bacteria-proof filter and it was shown that the resultant eluant was still infective. From this Beijerinck proposed the theory of *contagium vivum fluidum*, from which the present term *virus* is derived. The requirement for vectors to transmit viruses began to be realised in this period. Workers in Japan showed that dwarf disease of rice was spread by a leaf hopper (*Nephotettix nigropictus*), while in the USA it was demonstrated that a single leaf hopper (*Circulifer tenellus*) taken from a sugar beet plant infected with curly top and placed for five minutes on a healthy beet plant could transmit the organism causing disease. Following this there has been intensive work to isolate and identify viruses.

Viruses are a heterogeneous group of non-cellular, akaryotic, obligate and mainly necrotrophic pathogens. It is difficult to relate viruses to each other in any phylogenetic system although some tenuous groupings have been attempted. There are at least 600 plant pathogens that are thought to be viruses and have been named. Two-thirds of these have not been rigorously proved as viruses and might be other organisms. Two definitions may be cited to indicate the general nature of viruses: 'sub-microscopic infective entities that multiply only intracellularly and are potentially pathogenic', and slightly more precisely 'transmissible parasites whose nucleic acid genome is less than 3×10^8 daltons in molecular weight and which need ribosomes and other components from their host cells for multiplication'. A dalton is a unit of mass equal to the mass of one hydrogen atom. A limitation of 3×10^8 daltons is sufficiently large to include bacteriophages, poxviruses and iridoviruses. The latter definition also includes viroids, such as that causing potato spindle tuber, which consist of ribose nucleic acid (RNA) but have no coat protein. The necessity to include a requirement to invade living host cells in order to replicate as part of any definition of the virus state becomes lessened as more instances of virus synthesis *in vitro* are reported.

Virus particles cannot be viewed directly because of their small size (20–1200 nm) but by crystallographic techniques and electron microscopy four general shapes have been identified.

1. *Rods*, which are short, stiff and straight-sided, for example, tomato mosaic virus.
2. *Filaments*, which tend to be longer than rods but of smaller diameter with a sinuous shape, for example, potato virus X.
3. *Isometric particles*, which have a regular icosahedral shape, for example, turnip yellow mosaic virus and satellite virus.
4. *Bacilliform* or bullet-shaped, which have straight sides and rounded ends, for example, broccoli necrotic yellow virus.

Each virus particle (*virion*) is composed of nucleic acid which gives it an infective capacity. The nucleic acid is surrounded by a protein coat (*capsid*). The entire virion may or may not be enclosed in an envelope containing lipids or lipoproteins. When this envelope is absent the virion is termed *naked*. The protein shell provides protection when the virus is outside a host cell. The amount of nucleic acid appears to vary with particle shape; thus isometric particles contain 15–45 per cent, rod-shaped particles contain 5 per cent and bacilliform particles contain 1 per cent. The remainder is largely protein although lipid content may reach 20 per cent. Some viruses have been found to contain enzyme proteins. Up to 50 per cent of each particle is water but this is is generally disregarded when discussing virus composition. Additionally, various metal ions and up to 1 per cent polyamine may be present. The nucleic acids are composed of a long skeleton of alternating sugars (ribose or deoxyribose) with phosphate residues and purine or pyrimidine bases. The purines are adenine or guanine and the pyrimidines are cytosine, uracil (for those with ribose sugar) or thymine (for those with deoxyribose sugar). Until recently plant viruses were thought to contain solely ribonucleic acid (RNA) but Shepherd *et al.* (1968) showed that cauliflower mosaic virus (section 3.5.1) contains double-stranded deoxyribose nucleic acid (DNA). The sequence of bases in the nucleic acid and hence proportions of the four bases (*base ratio*) is a characteristic of each virus. Little information is available on the composition of plant virus proteins. Amino acid sequences in the coat proteins of tobacco mosaic virus (Fraenkel-Conrat, 1969) and turnip yellow mosaic virus (Stehelin *et al.*, 1973) have been determined. Multiplication of virus particles takes place by *replication*; in this process viral protein and nucleic acid moieties increase within susceptible host cells.

This is accomplished by taking over the replication machinery for host nucleic acid and protein synthesis and causing it to assemble virus particles rather than host protein and nucleic acids. More detailed information concerning virus structure and multiplication is to be found in: Kaper (1975), Smith (1977), Martin (1978) and Primrose and Dimmock (1980).

3.2 VIRUS TRANSMISSION

There are three principal modes of virus transmission.

Contact

This may take place in nature where plant parts contact each other, for example, when leaves of infected and healthy plants press together during winds, causing damage to the leaf edges and hairs thereby providing wounds for virus transmission; potato virus X is thought to spread in this manner. In woody species the roots of adjacent plants become grafted together and apple mosaic virus (section 3.5.6) may spread via such unions (Hunter *et al*., 1958). Contact spread can also be taken to include transmission by human agency as when tomato mosaic virus in sap is carried on the hands and knives of glasshouse staff. Transmission also occurs during vegetative propagation when infected material is multiplied or infected scions or rootstocks are used in conjunction with healthy plants. Many viruses are maintained by vegetative propagation apparently without any other means of spread.

Seed and pollen

In general, viruses are not transmitted in the sexual process but a minority of about 30 are capable of transmission either in the embryo or surrounding tissues or are transmitted passively with seed. This mode of spread provides foci for the further development of infection in succeeding crop generations, for example lettuce mosaic virus (section 3.5.2), or may adversely affect seed viability as with some pea viruses. Infection takes place before the ovules are fertilised and the level of infection depends on the relative times of infection and flowering and the position of flowers on the plant. Extremes of temperature will reduce the level of infection. Some host families, such as the Leguminosae, are more prone to seed infection and viruses of the Nepovirus type (section 3.4) are more frequently seed transmitted. The latter are important pathogens of perennial horticultural crops. Transmission via pollen grains (paternal source) is less effective than via ovules (maternal source). Viruses such as tomato mosaic virus, which spread in and on seed testas, can be inactivated by seed treatment with trisodium phosphate or hydrochloric acid but any residual inoculum may persist with stored seed for several years. The reasons why seed transmission is a restricted avenue for virus spread are unclear. Most viruses, however, spread within the phloem which is not contiguous with that of the embryo; consequently there is no direct route for invasion. Once inside the embryo, viruses may be unable to replicate or may be actively degraded by meristematic tissues.

Vectors

Many viruses require a mobile agent (*vector*) for spread. A range of plant-sucking and biting insects, mites, free-living soil-borne nematodes and soil-borne fungi will transmit virus particles (table 3.1). It is possible for several vectors to transmit the same virus; where this happens such vectors are usually closely related taxonomically. All vectors have one feature in common in that they penetrate unwounded plant cells, usually when feeding, thereby acquiring and transmitting virus particles. Biting insects such as beetles, flea beetles, earwigs, grasshoppers and weevils either retain a virus for periods up to 20 days (for example, turnip yellow mosaic virus) or for shorter periods of 1-5 days (for example, turnip rosette virus). These animals lack salivary glands and consequently transmission takes place during regurgitation from the stomach. With white flies, which are sucking insects, transmission efficiency increases with longer feeding periods at the virus source with a relatively short incubation period and the virus may be retained for up to 20 days by the vector. It takes longer for transmission of a virus by white flies than for them to acquire the pathogen because it is necessary to penetrate the phloem. Many leaf hopper-transmitted 'viruses' are now realised to be mycoplasmas (section 4.2) but with true viruses transmitted by leaf hoppers there is an intimate relationship between virus and vector with no latent period (for example, Tungo disease of rice). Lacewing vectors are capable of retaining virus particles of, for example, beet leaf curl virus through periods of hibernation by the adult. At least ten viruses are transmitted by Eriophyid mites (blackcurrant reversion is now thought to be caused by a mycoplasma); generally the adult stage is unable to acquire or transmit viruses *de novo*. Infection takes place during the nymphal stages and is carried over into the adult. The majority of viruses are spread by aphids; of 400 spp. of insects known to transmit some 200 different viruses, the aphid *Myzus persicae* carries over 78. A series of terms is used to describe the relationship between vector and virus. Although these terms can be applied to most vectors, they were largely initiated from studies with aphids.

1. *Acquisition access period*—the time during which a virus-free vector is permitted access to infected plant material.
2. *Acquisition feeding period*—the time during which a virus-free vector appears to feed on infected plant material.

Table 3.1

Examples of vectors of plant viruses and virus-like agents (Gibbs and Harrison, 1976)

Type of vector	*Approximate number of vector species*	*Approximate number of viruses transmitted**	*Vector: virus relationship***
Insecta			
Hemiptera			
Aphididae (aphids)	200	160	N, S, P
Psyllidae (psyllids)	1	1	—
Coccidae (mealy bugs)	15	2	S
Aleyrodidae (white flies)	3	25	(N), S, P
Auchenorrhyncha (leaf hoppers and plant hoppers)	60	35	S, P
Gymnocerata (tingid bugs)	2	2	P
Coleoptera (beetles)	30	20	S, P
Thysanoptera (thrips)	6	3	P
Orthoptera†	10	6	—
Dermaptera†	1	1	—
Lepidoptera (moths)†	4	5	—
Diptera (flies)†	2	3	—
Acarina			
Eriophyidae (leaf and bud mites	8	10	(S), P
Tetranychoidea (spider mites)	2	3	—
Nematoda (nematodes)			
Dorylaimida	20	14	P
Gastropoda†	6	1	—
Fungi			
Chytridiales	3	6	N, P
Plasmodiophorales	3	3	P

* = No vectors are known for more than 250 viruses and virus-like agents.
**N, viruses persist a few hours (non-persistent); S, persist 1–4 days (semi-persistent); P, persist weeks in the vector (persistent); ( ), tentative category; — no categorisation possible.
† = Vectors in these taxa are thought to have little pathological importance.

3. *Inoculation (or test) access period*—the time during which a virus-carrying vector is permitted access to uninfected plant material.
4. *Inoculation (or test) feeding period*—the time during which a virus-carrying vector appears to feed on infected plant material.
5. *Latent (or pre-infective) period*—the time between the beginning of an acquisition feeding period and the vector having the capability to

infect healthy plants; this may be quoted as the latent period for half the vector population (LP_{50}).

6. *Transmission threshold period*—the total time between acquisition of a virus by a vector and inoculation into a healthy plant.
7. *Persistence*—this relates to periods during which a vector is infective after leaving a virus source and is defined at three arbitrary levels.

a. *Non-persistent (originally external)* infectivity of the vector persists for a few hours (generally less than 4 h) at 20°C. Virus particles are quickly picked up by the vector on mouth-parts; probably in less than a minute. Non-persistence may result from inhibitory substances that are present on insect mouth-parts and gradually reduce virus stability. Specificity is not well developed in non-persistent viruses; one virus may be transmitted by several vector species. They are easily transmitted experimentally by sap inoculation and the short acquisition and inoculation periods indicate that the virus invades host epidermal cells. These viruses are not normally carried through moult periods by a vector.

b. *Semi-persistent* infectivity of the vector persists for 10-100 h but gradually declines with time; there is no latent period between acquisition and the ability to transmit virus particles and infectivity does not survive periods of moulting. These are essentially non-persistent viruses that persist for anomalously long periods and are acquired from and inoculated into deep-seated plant tissue such as the phloem. Generally, semi-persistent viruses are more specific in their vector requirement than non-persistent types.

c. *Persistent (originally internal)* infectivity of the vector persists for periods greater than 100 h, even for the whole vector life. It is often necessary for a vector to feed for an hour or more before the virus is acquired. Viruses of this type cause leaf-rolling and yellowing symptoms and the particles are concentrated in the phloem. They are not sap transmissible and are retained by aphid vectors through periods of moulting. The terms persistent and non-persistent were coined by Watson and Roberts (1939).

Once inside a vector, viruses may be described as follows.

a. *Circulative*, which pass from an insect gut into its circulatory system and back to its salivary glands. Once this cycle is accomplished the vector is infective. Inevitably there is a delay or *latent period* between acquisition of a virus and the ability of a vector to transmit it.

b. *Propagative*, which are thought to multiply within a vector. Evidence is not strong since it is necessary to prove that there is an increase in particle number following acquisition by the vector and that this increase is not merely a result of further acquisition.

c. *Stylet-borne* which are carried solely on vector stylets. These terms were coined by Kennedy *et al*. (1962).

These definitions may be equated approximately as

Non-persistent	Stylet-borne
Semi-persistent, Persistent	Circulative or propagative

Viruses that are spread by free-living nematodes are of the following two types.

Nepoviruses

These have a wide host range and induce ringspot symptoms. The vectors are large ectoparasitic nematodes such as *Longidorus* spp. and *Xiphinema* spp. These possess a feeding apparatus consisting of a tubular stylet leading into an oesophagus (gullet) the posterior part of which possesses a thick muscular wall where salivary glands are located. Some *Xiphinema* spp. feed mainly but not exclusively at root tips, inducing small galls; others feed on the sides of young roots, causing cortical necrosis and destroying feeder roots. *Longidorus* spp. feed mainly at root tips, arresting growth and inducing clubbing. There is considerable specificity between nepoviruses and their vectors. *Xiphinema*-spread viruses persist in a vector for longer than those spread by *Longidorus* spp.; for example raspberry ringspot virus (section 3.5.4) and tomato blackring virus persist in *L. elongatus* for less than 3 months whereas tobacco ringspot virus persists nearly 12 months in *X. americanum*. There is no evidence of multiplication within a vector and viruses are not retained through periods of moulting. Most nepoviruses are pathogens of endemic wild plants; hence weed hosts serve as virus reservoirs. Elder (*Sambucus nigra*) and blackthorn (*Prunus spinosa*) harbour Arabis mosaic virus and chickweed (*Stellaria media*) harbours raspberry ringspot virus (section 3.5.4). Seed and pollen transmission are important modes of spread for nepoviruses, more so than other groups of nematode-transmitted viruses. This varies with different viruses. It is less common with Arabis mosaic virus and strawberry latent ringspot virus which probably persist through a winter in the vector *X. diversicaudatum* and more common with raspberry ringspot virus (section 3.5.4) and tomato blackring virus which seem to be more frequently lost by *L. elongatus* during winter. Survival and spread of these viruses at a particular site depends on the occurrence and abundance of infected weeds and weed seeds, the persistence of a virus in, and numbers of, vector nematodes and the effects of soil temperature and moisture content on nematode movement and feeding.

Tobraviruses

These include tobacco rattle virus which causes diseases of various bulbous ornamentals and pea early browning virus, a major pathogen of vining pea crops. The vectors are ectoparasitic nematodes such as *Trichodorus* spp. and *Paratrichodorus* spp. These nematodes feed on young roots in the root hair region and the zone of elongation causing browning and inhibition of root growth, resulting in 'stubby-root' symptoms. Particles of both tobraviruses and nepoviruses are acquired during the feeding process but are transmitted successfully only to root cells which suffer slight damage so that their metabolic processes are unimpaired. There is no evidence for a latent period following acquisition of tobraviruses and they may persist for months or years in a nematode. The particles become associated with the cuticular lining of the pharynx and oesophagus and are probably ejected with saliva during feeding. Essential features in the tobravirus–vector–host plant relationships are: the wide virus host range, gregarious feeding habits of the vector nematodes, the ability of vectors to acquire and inoculate virus particles quickly and the long persistence of infective virus particles within the vector. Geographical distribution of vector nematodes is related to the occurrence of soil series; they are predominantly found in sandy loam to loamy sand soil types. Weeds such as chickweed (*Stellaria media*) are reservoirs of virus and seed transmission takes place in some host species such as field pansy (*Viola tricolor*). Movement of nematodes is encouraged by adequate soil moisture and temperatures of 15–20°C. Virus survival depends on the presence of infected weed hosts and weed seeds; persistence is limited in the vectors which are easily damaged and susceptible to desiccation.

Fungal vectors

Viruses with fungal vectors include cucumber necrosis virus and some types of tobacco necrosis virus. They are usually confined to the roots of naturally infected hosts. Tobacco necrosis virus has a wide range causing such diseases as the necrosis of tulip shoot tips (Augusta disease). It is transmitted by the uniflagellate zoospores of Chytridiomycete fungi such as *Olpidium* spp. which are obligate parasites colonising the roots of a wide host range. Virus particles are thought to occur *externally* on resting spores and zoospore flagella. The mode of infection is not known but the virus probably survives away from the vector in living roots, root debris, soil water and adsorbed to clay particles in soil. Some viruses are carried *internally* in fungal resting spores. These include potato mop-top virus and beet necrotic yellows virus whose vectors are fungi such as *Spongospora* spp. and *Polymyxa* spp. Soil moisture content is a major factor influencing the spread of, for example, potato mop-top virus.

Transmission of viruses results from the successful conclusion of four events: acquisition of virus particles by a vector, survival of infectivity during the association of virus and vector, inoculation into the host plant at the correct place (for example, epidermal or phloem tissue) and subsequent initiation of infection (figure 3.1). Control measures may be initiated at any one of these points (section 9.2). Transmission may be more complicated where disease apparently results only from the presence of two unrelated viruses. In this situation there may be an interaction between two viruses so that one acts as a *helper* for transmission of the other. It is possible that the helper virus supplies an as yet unidentified virus-coded protein which plays a role in vector transmission. An example of this relationship in horticultural crops is carrot mottle virus for which the principal vector is the aphid *Cavariella aegopodii* but this virus cannot be transmitted without the presence of carrot red leaf virus which is also spread by *C. aegopodii*. Together these viruses give rise to carrot motley dwarf syndrome.

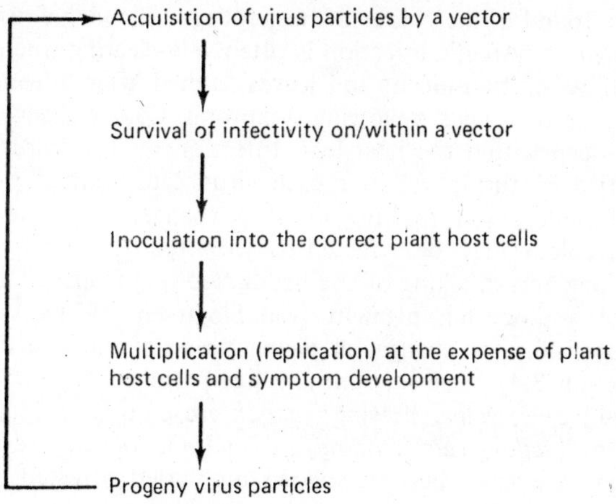

Figure 3.1 Infection cycle of virus pathogens transmitted by vectors

3.3 EFFECTS OF VIRUS INFECTION ON HOST PLANTS

Infection by plant viruses takes place through wounds; this contrasts with animal viruses which can invade unwounded cells. Once a plant is infected it usually remains so for the rest of its life; hence survival can in some instances be for very long periods especially where the host is multiplied by vegetative means. Typically, all plant tissues with the possible exception of meristems in roots and shoots are invaded. This leads to a wide range of abnormalities in host plants. Chlorosis and necrosis are generally associated with virus infection in a similar way to infections by fungi and bacteria. Symptoms develop in a sequential manner, often beginning as small dead zones of tissue on those leaves that were infected first; these are termed *local* or *primary lesions*. Such lesions may expand showing areas of chlorosis surrounded by concentric rings of dead tissue (*necrotic ringspots*). Symptoms may then develop on previously uninfected foliage and are said to have become *systemic*. Most viruses cause forms of abnormal growth pattern in a host by restricting or accelerating growth. This may result in alterations to leaf morphology as is found in tomatoes infected with cucumber mosaic virus where the leaves develop with little or no lamina tissue giving rise to *fernleaf* or *shoestring* symptoms.

An example of increased growth is seen when raspberry ringspot virus (section 3.5.4) infects cherry and induces leafy outgrowths (*enations*) on the undersurfaces of leaves. Flowering and seed production may cease, as in broad beans infected by bean leaf roll virus where the flowers abscise and any ovules fertilised prior to infection form seeds of reduced size. Changes are found in the colour of host foliage following virus infection; an initial sign of systemic infection is often vein-clearing where the veins become yellow or translucent and leaves formed after infection exhibit mosaic, mottle or general yellowing symptoms. Colour changes may also take place deep within the host leaf. Infection by Tymoviruses (section (3.4), typified by turnip yellow mosaic virus, cause clumping and degradation of the chloroplasts resulting in colour changes within the leaf.

Flower-colour may be affected as with tulip breaking virus which causes flecking and streaking of the petals resulting from variations in the amounts of anthocyanin pigment (van Slogteren, 1971). Viruses may also be responsible for changes in pigmentation of the seed testa. Nepoviruses (section 3.4) cause ringspot and line-pattern symptoms on leaves. Other viruses cause vein-yellowing (*vein chlorosis*) or yellowing on either side of veins (*yellow vein banding*). Virus infection may result in leaf-rolling where the leaves become cupped and curled, affected tissue being turgid and brittle.

The effects of virus infection are not restricted to those immediately obvious to the eye; host cytology, histology and physiology may be affected. Tumours and enations are manifestations resulting from increased cell division. Viruses affect the vascular system; some cause changes reminiscent of fungal pathogens (section 2.6.9) with tylosis formation in xylem vessels. Where the phloem is invaded, cells die or degenerate and callose deposits accumulate on sieve plates. At the subcellular level tymoviruses (section 3.4) cause vesicles to form within chloroplasts and mitochondria which become modified aggregating into *inclusion bodies*, for example, when plants are infected by tobacco rattle virus. The predominant characteristic of all infected cells is the presence of virus particles.

These are usually found in the cytoplasm but may occur in chloroplasts, spaces between the inner and outer nuclear membranes (*perinuclear space*) and within nuclei. Virus particles may be scattered or aggregated in various ways: spherical particles of tomato spotted wilt virus are found in membranous bags; elongated particles of beet yellow virus develop in bundles which aggregate end to end forming banded inclusions, while those of tobacco rattle virus are packed around the mitochondria. Various larger cytoplasmic inclusions are found, which may or may not contain virus particles. Those most exclusively viral are, for example, the crystalline plates found in tobacco plants infected by the type strain of tobacco mosaic virus or the more rounded plates in hosts infected by the ribgrass strain of this pathogen. Although the form of inclusion is a characteristic of invasion by a particular virus, a single virus type may be associated with several different forms. For example, the tomato Aucuba strain of tobacco mosaic virus is associated with hexagonal crystalline plates, spindle-shaped paracrystals and flat angular layers of aggregates. Some inclusions contain material other than virus particles. Although the amorphous 'X' bodies associated with tobacco mosaic-infected cells contain some virus particles, they are mainly a mixture of host cell components and flexuous tubules of unknown composition (Esau and Cronshaw, 1967). Pinwheel inclusions are characteristic of cells infected by many of the flexuous potyvirus types (section 3.4) with particles of 700–900 nm. These inclusions consist of proteins characteristic of the virus but serologically unrelated to that of the virus particle. Distribution of inclusions is related to that of virus particles within the host. Within infected cells there is increased cytoplasmic streaming and clumping of organelles, and, in those in close proximity to a developing lesion, the nucleoli are often swollen.

Spectacular effects on host physiology have been reported for plants infected by all types of pathogen—not least by viruses. It is often difficult to confirm a *causal relationship* since changes in physiology may not be directly caused by pathogen infection and physiological changes may also result from effects such as nutrient imbalance or tissue senescence consequent upon infection. Nonetheless there is now a substantial body of evidence to support claims for significant changes in host physiological processes following virus invasion. There are increases in host alcohol-soluble amino, amide and ammonia nitrogen, RNA and DNA phosphorus. Although total nitrogen and protein nitrogen content of tobacco leaves is little affected by infection with mosaic virus, virus proteins account for 75 per cent of total protein so that there is a considerable change in distribution apparently at the expense of normal host proteins. Temporary effects are found, such as a diversion of products from photosynthesis into organic and amino acids rather than sugars. Rates of translocation are affected by virus infection. Glucose, fructose and sucrose accumulate in infected leaves not because of increased synthesis but as a result of lower rates of translocation from leaves to root tissue. Respiration rates are affected, often within hours of invasion, but only increase dramatically just prior to symptom formation. Consequently, large changes in respiration are associated with plant–virus combinations where symptom development is severe; where symptoms are slight, only small shifts in respiration rate are detected. Enzyme systems in virus-infected cells have been shown to change quantitatively but, so far, qualitative changes have not been proved, although it is thought that some RNA polymerases found in cucumber plants infected by cucumber mosaic virus may be virus coded.

There is apparently considerable difference between the ways in which viruses, as opposed to other micro-organisms, multiply (*replicate*) within host cells. Bacteria, for instance, within a host cell remain intact and recognisable at all stages of division (section 4.1). Once a virus enters a host there is partial or complete disassembly into constituent macromolecules. Then the virus genome (nucleic acid carrying genetic information), or messenger RNA transcribed from it, is translated via host-translation apparatus into new virus protein. The nucleic acid genome is replicated and progeny nucleic acid together with molecules of one or more new virus proteins assemble forming progeny virus particles. Consequently, intact virus particles are not present during part of the replication cycle; the virus genome is incorporated into host protoplasm and is totally dependent upon it. This process begins immediately entry of a virus has been effected through ruptured host cell walls either through mechanical damage or by a vector piercing the cell.

Virus particles move from cell to cell via the plasmodesmata and from one plant organ to another via elements of the vascular system, principally in the phloem. They rarely enter mature leaves or meristematic tissue. Virus particle concentration is dependent both on the virus and the type of host. For instance the concentration of tobacco mosaic virus in tobacco is five to ten times greater than in tomatoes. Concentration is also dependent on environmental conditions and the length of time over which plants have been infected. Highest concentrations are usually found in the first leaves to be systemically infected. Nepoviruses (section 3.4) cause bright ringspots in the earliest infected leaves but those leaves invaded later are virtually symptomless and these differences are correlated with the concentration of virus particles in the respective leaves (Price, 1936). It is also noted that diseases of a mosaic type pass through cycles of severe and mild symptoms which are directly related to changes in virus particle concentration (Paul and Quantz, 1959). Although symptom severity is correlated with virus concentration for a particular host–virus combination, the relative severity of symptoms caused by two distinct viruses in the same host or by a single virus in two different hosts may not indicate the relative concentrations of virus particles.

Environmental conditions such as temperature, host nutrition, intensity and duration of light and availability of water all affect virus concentration which is often highest in the most vigorously growing hosts. For instance, plants receiving adequate nitrogen and phosphate fertilisers were found by Bawden and Kassanis (1950) to contain 40 times more tobacco mosaic virus than deficient hosts. Temperature will significantly affect virus content, with optima influenced both by host and virus requirements and differing for a range of viruses in the same host. It is noted that viruses can be inactivated by exposure of a host to elevated temperatures (sections 3.5.6 and 9.2). Infectivity falls more quickly than the concentration of virus particles which suggests that particles can be inactivated without losing their protein coat.

3.4 VIRUS CLASSIFICATION

Classification and grouping of viruses is essential in order that they may be discussed and described adequately. It is of fundamental practical importance that newly found viruses should be related to previously described individuals or groups since by analogy much may be deduced about their possible behaviour, such as prediction of likely vectors and field ecology and hence means of limiting spread. This leads to more direct and reliable research to obtain methods of control. Initially viruses were named according to the first host plant on which they were discovered and then given a number, for example, references are found in the older literature to cauliflower virus 1, cucumber virus 1 and *Hyoscyamus* virus 3. Thereafter a Latinised binomial system was proposed but failed to achieve any widespread adoption since it was not based on natural principles. Most virologists use vernacular names for the entities that they study, for example tomato mosaic virus (TMV). Until recently the characteristics of these entities have been described on a qualitative basis using attributes such as the following.

1. *Host range*—both of crops and weed types.
2. *Symptom types*—such as the form of lesions found on inoculated leaves (local lesions) or those following systemic virus invasion (systemic symptoms).
3. *Spread by vectors*—types of vector concerned, persistence of the virus in a vector and host plant.
4. *Means of artificial infection*—transmission in expressed sap or by artificially introduced vectors.
5. *Infectivity of serially diluted sap*—giving the dilution-end-point, that is, the point at which virus concentration is so low that no infection can be detected.

6. *Stability of infective sap*—this is affected by elevated temperatures and various chemicals.
7. *Serological tests*—when plant viruses are injected into mammals such as rabbit an antibody reaction is established; serum prepared from the mammalian blood is specific against the originally injected virus and will cause a precipitation reaction when the two are mixed. This is a very specific reaction and can be used to identify individual viruses and often strains of a particular virus.
8. *Interference reactions*—plants infected by one virus strain are sometimes protected against infection by other strains; this technique can be used to determine some virus relationships.
9. *Sedimentation rate*—virus particles of different types placed in a density gradient of sucrose and centrifuged at high speed (*ultracentrifugation*) sediment at rates dependent on their mass; this technique is useful to separate virus particles of distinct mass which may be identified further by other means.

Early in the study of viruses it was realised that they behave in a similar manner to other living organisms in that progeny differ from their parents. Genetic variants are formed when viruses multiply; these may induce contrasting symptoms and may differ in their host range, ecology and vector specificities (Jensen, 1933). Consequently, strains and variants are identified within a single virus type. Some of the characteristics that may be used to distinguish between strains and variants are shown in table 3.2. These properties are of two types.

1. *Those that may be measured in absolute terms*, for example, gross composition and particle dimensions.
2. *Those of a qualitative nature*, which compare the properties of viruses by their reaction with host plants and animal vectors.

From these approaches has developed a more systematic form of virus classification whereby the vernacular name is made more definitive by adding a *cryptogram* which in a coded form presents some of the more important properties of a virus. The major characters used in these cryptograms are shown in table 3.3. Using this system, viruses with similar properties may be grouped together around a *type* (sum of characteristics common to a large number of individuals and serving as a ground for classification) and from a prominent member of each group a general name is derived. This system is illustrated in table 3.4.

Table 3.2

Characters used to differentiate between virus isolates (Gibbs and Harrison, 1976)

1. *Unrelated Viruses*—these usually differ by one or more genetically stable characters such as:
   - 1.1 Type, strandedness and number of pieces of their nucleic acid genome
   - 1.2 Size, shape, symmetry and gross composition of their particles
   - 1.3 Size and number of polypeptides in their particles
   - 1.4 Hydrodynamic behaviour of their particles
   - 1.5 Ability of their particles to react with one another's antisera
   - 1.6 Ability of their particles to be photo-reactivated after ultraviolet irradiation
   - 1.7 Type of symptoms induced in host plants
   - 1.8 Taxa of their natural vectors
   - 1.9 Type of reaction with their vectors

2. *Groups of Viruses*—these have in common most, if not all, of the characters in (1) but may:
   - 2.1 Differ in the amino acid composition of their coat proteins so that their particles have different electrophoretic mobilities and are serologically only distantly related or unrelated
   - 2.2 Have different, but related, vector species
   - 2.3 Differ in their host range and symptom severity though still causing similar types of symptoms in host plants

3. *Strains of Viruses* (or variant progeny of a single virus)—these normally have the majority of their properties in common but may:
   - 3.1 Differ slightly in the amino acid composition of their coat proteins and thereby differ slightly in serological specificity, electrophoretic mobility or particle stability
   - 3.2 Differ in vector species specificity or the ease of transmission by vectors
   - 3.3 Cause symptoms of differing severity

3.5 EXAMPLES OF PLANT-PATHOGENIC VIRUSES AND THEIR MODES OF SPREAD

Since no natural classification is available for viruses, the six examples given here are treated according to their main mode of transmission because control is exercised through a knowledge of the mode by which viruses are spread.

3.5.1 Cauliflower mosaic virus (CaMV)—spread by insects (aphids)

The following aphids which breed on brassicas are vectors of CaMV: *Brevicoryne brassicae* (cabbage aphid), *Lipaphis erysimi* (false cabbage aphid) and *Myzus persicae* (peach aphid). A range of other aphids that do not breed on brassicas have been shown to transmit CaMV under artificial conditions.

Table 3.3

Characters used to classify viruses by means of a coded cryptogram (Gibbs et al., 1966).

1. Type of nucleic acid and degree of strandedness
2. Molecular weight of nucleic acid (in millions) and percentage of nucleic acid in the infective particle
3. Outline of the particle and outline of the nucleocapsid
4. Kind of host infected and kind of vector

Additional symbols

* = Property of the virus is not known
( ) = Enclosed information is doubtful or unconfirmed
[ ] = Enclosed cryptogram gives information about a virus group

Example

cucumber mosaic virus (CMV) = $\frac{R}{1} : \frac{1}{18} : \frac{S}{S} : \frac{S}{Ap}$
(1) (2) (3) (4)

(1) $\frac{R}{1}$ — R = ribonucleic acid; 1 = single stranded

(2) $\frac{1}{18}$ — 1 = molecular weight; 18 = percentage of nucleic acid in infective particles

(3) $\frac{S}{S}$ — S = spherical; S = spherical

(4) $\frac{S}{Ap}$ — S = seed plant; Ap = Aphididae

Symptoms

Symptoms develop before and after transplanting when autumn and winter cauliflowers have been infected in the seed-bed. Where infection takes place after transplanting, symptoms may not appear until the plants are semi-mature. Affected leaves show a mottling pattern of light and dark green with frequently a very dark green band up to 5 mm wide adjacent to the major veins. Infected plants are often stunted with the apical leaves being smaller than in healthy plants. Curd size and quality are reduced and the wrapper leaves are too small to protect the curd from frost. Winter hardiness is reduced and, in consequence, plant population is diminished. Foliar symptoms on sprouting broccoli, summer cauliflower, Brussels sprout and kale tend to be less pronounced than on winter cauliflower. Even so, under conditions of severe infection Brussels sprout leaves develop extensive mottling often with a yellowish vein-banding, while infected kale leaves show yellowish and dark-green vein-banding.

Table 3.4

Group names and type members used in the virus classification scheme proposed by Gibbs and Harrison (1976)

1. *Viruses with helically constructed tubular or filamentous particles*

	Virus group	*Type member*
1.1	Tobravirus	Tobacco rattle virus
1.2	Soil-borne wheat mosaic and potato mop-top virus	
1.3	Tobamovirus	Tobacco mosaic virus
1.4	Potexvirus	Potato virus X
1.5	Carlavirus	Carnation latent virus
1.6	Potyvirus	Potato virus Y
1.7	Beet yellows virus and citrus tristeza virus	

2. *Viruses with isometric particles*

2.1	Cucumovirus	Cucumber mosaic virus
2.2	Tymovirus	Turnip yellow mosaic virus
2.3	Comovirus	Cowpea mosaic virus
2.4	Nepovirus	Viruses with nematode vectors and polyhedral particles
2.5	Tobacco necrosis virus	
2.6	Satellite virus	
2.7	Bromovirus	Brome mosaic virus
2.8	Tombusvirus	Tomato bushy stunt virus
2.9	Potato leaf roll and similar viruses	
2.10	Viruses with labile particles of two or more types	
2.11	Pea enation mosaic virus	
2.12	Caulimovirus	Cauliflower mosaic virus
2.13	Clover wound tumour and similar viruses	
2.14	Tomato spotted wilt virus	

3. *Viruses with bacilliform or bullet-shaped particles*

3.1	Alfalfa mosaic virus	
3.2	Cacao swollen shoot virus	
3.3	Rhabdovirus group	

4 *Viroids*

These are virus-like pathogens that do not produce characteristic nucleoprotein particles; generally they are transmitted mechanically and have no confirmed animal vectors. Examples are potato spindle tuber disease and chrysanthemum stunt

Control

Control can only be achieved by keeping newly sown crops well separated from old established and possibly infected ones. The residues from crops should be destroyed by ploughing, burning or feeding to stock immediately harvesting is finished. Where seedlings are raised on the same farm as mature plants they should be well separated; the seed-beds should be kept away from hedges and preferably surrounded by cereals to prevent aphid vectors reaching the young plants. Aphicidal sprays will not prevent infection once the vector has landed and begun to feed on the cauliflower since infection occurs very rapidly. It is preferable in areas where there is a high risk of infection to buy in transplants that have been raised in areas where cauliflower and other brassica crops are not grown. Weed control is essential since several wild crucifers are host to this virus.

3.5.2 Lettuce mosaic virus (LMV)—spread by seed

Although LMV can be transmitted locally by aphid vectors its world-wide distribution is due to spread by infected seed. Transmission into developing seed occurs through both the pollen and the ovules of infected plants. The level of seed infection ranges from 1–10 per cent depending on the age of the seed mother plant and cultivar. Seed transmission does not take place in cv. Cheshunt Early Giant, for example, because the primary flower heads are killed by LMV and later-formed flowers produce virus-free seed.

Symptoms

Symptoms on mature butterhead lettuce consist of dwarfing, malformed hearts, mottling, yellowing, necrosis and leaf distortion. Mottling and yellowing are more prominent in the early season whereas necrosis and scorching become evident in summer. Vein-clearing is common in both infected seedlings and mature plants together with irregular pale blotching and the whole leaf may take on a hard yellow appearance similar to effects caused by wind damage. Leaf morphology is changed with increased blistering and ballooning while the marginal serrations typical of some cultivars disappear when they are infected by LMV. Necrosis begins as small lesions between the veins or as definite veinal necrosis and marginal leaf scorch. Mosaic becomes evident when the flower stalk is produced; the leaves on the stem and inflorescence exhibit pronounced mottling possibly accompanied by necrosis and the whole plant may be stunted. In some cultivars necrotic lesions form on the stem, bracts and flower buds while the flowers may fail to open. Seed production is reduced. Once the host lettuce plant is infected, the virus is distributed systemically within it. Vein-clearing is a primary symptom on cos lettuce cultivars with a more or less intense mottling. Infected plants tend to be stunted with leaf-scorching but veinal chlorosis is less frequent than in butterhead types. Failure to produce a dense compact heart is a useful diagnostic symptom if mottling is not expressed.

In addition to various weed hosts, LMV will also infect *Aster*, *Cineraria*, endive, garden pea, mangold, sweet pea, *Tagetes erecta* and *Zinnia elegans*.

Control

Control is primarily achieved by use of virus-free seed. Reputable seed-houses produce seed-crops in areas where the risk of infection is minimal and then *Index* the seed before sale. Indexing is done by growing samples of seed and then visually examining the seedlings for signs of infection. A seed-stock with contamination greater than 0.1 per cent constitutes a considerable economic risk to the grower. Although this would give only about 200 primary foci of LMV infection per hectare, subsequent spread by aphid vectors could devastate the crop. Similarly it is important to ensure that weed hosts of LMV, for example groundsel (*Senecio vulgaris*) and prickly sow thistle (*Sonchus asper*), are eradicated by cultivation and by use of herbicides before sowing or planting a lettuce crop. New drillings or plantings should not adjoin older crops which may be infected; crops of similar age should not be scattered about a holding but planted in blocks and separated from other crops spatially. Open sites are preferable to enclosed ones and seed-beds should be away from hedges, woods and buildings. Recently some Dutch lettuce cultivars have been developed with 'tolerance' to LMV so that, although they become infected, symptoms do not develop.

3.5.3 Lettuce big vein (LBV) organism—spread by a fungus

Several viruses are known to be transmitted by soil-borne fungi and although the causal organism of lettuce big vein has not been isolated and characterised it is known to be related to the invasion of lettuce roots by the fungus *Olpidium brassicae*. The causal agent of LBV is thought to be transmitted in the fungal zoospores. Soil containing *O. brassicae* carrying the agent of LBV can be air dried and retain infectivity for at least eight years.

Symptoms

Symptoms include a very marked vein-clearing followed by enlargement and bleaching of the vascular regions of petioles and leaf blades. Further vein-banding and pronounced savoying of the leaves develop as the host matures until infected lettuce plants have a stark appearance with whitened veins standing out against the greener interveinal tissue.

Control

Control of LBV for lettuce crops grown in soil can be achieved by steam sterilisation to eradicate the fungal vector. As alternatives to this expensive technique, chemical sterilants such as chloropicrin and DD mixtures have been advocated. Variably effective results have been obtained with these materials and their efficacy may be related to soil type. This pathogen is particularly troublesome in lettuce crops grown by nutrient film technique; in such systems a very low level of initial infection can rapidly build up to a devastating epidemic due to circulation of vector zoospores within the nutrient flow. Addition of fungicides and wetting agents to the system will combat such infections.

3.5.4 Raspberry ringspot virus (RRV)—spread by nematodes

Vectors of RRV are the free-living soil-borne nematodes *Longidorus elongatus* and *L. macrosoma*. This virus can also be transmitted in the seed of weed hosts such as chickweed (*Stellaria media*).

Symptoms

Symptoms include the development of interveinal chlorotic and necrotic zones, foliar puckering, curling and yellow blotching especially in the raspberry cv. Lloyd George. In both raspberry and strawberry, infection reduces the vigour of plantations leading to localised death of plants, reflecting the uneven distribution of the vectors, which migrate only slowly in soil. The raspberry cvs Norfolk Giant and Malling Orion in particular exhibit leaf curling symptoms; leaves become brittle and abscise under slight pressure. RRV is lethal to cvs Norfolk Giant, Malling Jewel and Malling Orion; Glen Clova is also susceptible. This virus also causes spoonleaf in redcurrant and leaf distortion of gooseberry. It is capable of infecting bean, cowpea, cucumber, grape vine, *Petunia* and tomato as well as a range of weed hosts. Most hosts probably transmit RRV through seed which in addition to aiding virus dissemination also provides a means of survival in soil.

Control

Control has been advocated through the elimination of vectors and by crop rotation. Although soil fumigants such as D-D mixture or PCNB will remove a high proportion of nematode vectors, some will remain to re-establish a population. Additionally the virus will still exist in weed seeds. Crop rotation is a sensible control because vector numbers decline in raspberry plantations which lessens the rate of infection of subsequent crops. In strawberry plantations vector numbers increase; hence disease is more likely to develop in raspberries planted after strawberries than in strawberries planted after raspberries. Control has also been advocated by use of 'immune' cultivars, none of which are commercially acceptable for agronomic reasons.

3.5.5 Carnation Mottle Virus (CarMV)—spread by manual handling

This virus is found widely distributed in the commercial florists' carnation. There is no known vector but it is thought to be readily transmitted in sap passed from plant to plant during husbandry and harvesting operations and by root contact between plants.

Symptoms

Symptoms are a faint or mild leaf-mottling, but in some cultivars no symptoms at all may be expressed. The waxiness of mature leaves and colouration of plants grown 'hard' can obscure symptoms. Flowers show faint streaking chiefly on the back of the outermost petals. There is a general loss of vigour in infected plants which is usually detected by comparison with healthy ones. Other *Dianthus spp*. such as *D. barbatus* (sweet william), *D. chinensis* and *D. superba* are susceptible to CarMV although symptom development may vary from slight chlorosis to a fleck mottling and leaf distortion. Symptoms may often fail to develop especially in *D. barbatus*. Some symptoms have been recorded in *Calandrinia grandiflora, Celosia argentea*, beet, spinach and tomato following artificial infection.

Control

Control in carnation has been achieved by meristem tip culture. Essentially this technique involves excising the apical dome, 0.2 to 1 mm long, from the tip of an actively growing shoot. This region of tissue is usually free from virus particles. The tissue dome is then transferred to a sterile tube containing a filter-paper bridge dipping into liquid nutrient. After 20-30 days the tip will have formed roots and can be transferred to sand or gravel irrigated with nutrient. Thereafter further virus-free material may be obtained and a nucleus of clean planting material established.

3.5.6 Apple mosaic virus (AMV)—spread by vegetative propagation

There are no known animal vectors of AMV which was one of the first plant 'variegations' to be shown to be transmitted through vegetative propagation techniques such as grafting and budding.

Symptoms

Symptoms caused by AMV on apple consist of various types of mottle with most commonly small irregularly shaped cream-to-yellow spots that are conspicuous against the background of healthy green leaf tissue. Such spots often coalesce to form large chlorotic zones. Infected leaves eventually become chlorotic. Chlorotic vein-banding may sometimes develop or there may be light-green and dark-green mosaic or areas of yellowish white tissue on the foliage. Most frequently only a few leaves will exhibit symptoms but occasionally whole shoots or boughs may be affected. Fruits do not show symptoms. Several strains of AMV have been identified but no specific symptoms can be associated with individual strains. By comparing symptom intensity in different apple cultivars it is possible to differentiate severe apple mosaic, vein-banding apple mosaic and mild apple mosaic strains. The definitions used to separate AMV strains are as follows.

1. White or pale-yellow flecks generally on a veinlet but without any general relation to the veins.

2. Large white or pale-yellow areas extending over two or more interveinal areas separated by main veins.
3. Vein-banding where both primary and secondary veins are bounded by white or pale-yellow zones.
4. Necrotic areas which often develop on the chlorotic lesions of 2 above.

Host range of AMV includes many important horticultural genera that are propagated by vegetative means, for example, *Chaenomeles*, *Crataegus*, *Cydonia*, *Fragaria*, *Malus*, *Prunus*, *Pyrus*, *Rosa* and *Sorbus* and the species *Acer campestris*, *Betula pendula*, *Cotoneaster bullata* and *Ligustrum ovalifolium*. Serological tests have shown that rose mosaic and apple mosaic are caused by the same virus.

Control

Control of AMV may be achieved by heat inactivation of the virus. Infected material is maintained at 36°C for between 3 and 10 weeks, depending on species. After this time a proportion of the material will fail to develop further symptoms. This is tested for freedom from virus by budding heat-treated material on to susceptible test material. Once virus-free material has been obtained it is propagated to commercial quantities in isolation and then distributed through trade organisations such as nucleus stock associations.

REFERENCES

BAWDEN, F. C. and KASSANIS, B. (1950). Some effects of host-plant nutrition on the multiplication of viruses. *Annals Applied Biology*, 37, 215–28

ESAU, K. and CRONSHAW, J. (1967). Relation of tobacco mosaic virus to the host cells. *Journal Cell Biology*, 33, 665–78

FRAENKEL-CONRAT, H. (1969). *The Chemistry and Biology of Viruses*, Academic Press, New York

GIBBS, A. J. and HARRISON, B. D. (1976). *Plant Virology–The Principles*, Edward Arnold, London

GIBBS, A. J., HARRISON, B. D., WATSON, D. H. and WILDY, P. (1966). What's in a virus name? *Nature, London*, 209, 450–54

HUNTER, J. A., CHAMBERLAIN, E. E. and ATKINSON, J. D. (1958). Note on transmission of apple mosaic by natural root grafting. *New Zealand Journal Agricultural Research*, 1, 80–82

JENSEN, J. H. (1933). Isolation of yellow-mosaic viruses from plants infected with tobacco mosaic. *Phytopathology*, 23, 964–74

KAPER, J. M. (1975). *The Chemical Basis of Virus Structure*, North-Holland Publishing Company, Amsterdam

KENNEDY, J. S., DAY, M. F. and EASTROP, V. F. (1962). *A conspectus of aphids as vectors of plant viruses*, Commonwealth Institute of Entomology, London pp. 1–114

MARTIN, S. J. (1978). *The Biochemistry of Viruses*, Cambridge University Press, London

PAUL, H. L. and QUANTZ, L. (1959). Uber den Wechsel der Konzentration des echten Ackerbohnen Mosaik Virus in Ackerbohnen. *Archiv. für Mikrobiologie*, 32, 312-18

PRICE, W. C. (1936). Virus concentration in relation to acquired immunity from tobacco ring spot. *Phytopathology*, 26, 503-29

PRIMROSE, S. B. and DIMMOCK, N. J. (1980). *Introduction to Modern Virology*, Blackwell, Oxford

SHEPHERD, R. J., WAKEMAN, R. J. and ROMANKO, R. R. (1968). DNA in Cauliflower Mosaic Virus. *Virology*, 36, 150-52

SMITH, K. M. (1977). *Plant Viruses*, Chapman and Hall, Science Paperbacks, London

STEHELIN, D., PETER, R. and DURANTON, H. (1973). Étude de la structure primaire de la proteine du virus de la mosaique jaune du navet. III. Séquéace des peptides chrymotrypsiques de la proteine aminoethylée et enchaînement des peptides trypsiques. *Biochimica Biophysica Acta*, 317, 253-65

VAN SLOGTEREN, D. H. M. (1971). Tulip Breaking. Commonwealth Mycological Institute/Association of Applied Biologists, *Descriptions of Plant Viruses*, No. 71. Commonwealth Mycological Institute, Kew

WATSON, M. A. and ROBERTS, F. M. (1939). A comparative study of the transmission of *Hyoscyamus* virus 3, potato virus Y and cucumber virus 1 by the vectors *Myzus persicae* (Sulz.), *M. circumflexus* (Buckton) and *Macrosiphon gei* (Koch.). *Proceedings Royal Society*, B127, 543-76

FURTHER READING

Commonwealth Mycological Institute/Association of Applied Biologists Descriptions of Plant Viruses, published by Commonwealth Mycological Institute Kew. Numbers: Apple Mosaic Virus 83; Carnation Mottle Virus 7; Cauliflower Mosaic Virus 24, 243; Lettuce Mosaic Virus 9; Raspberry Ringspot Virus 6, 198

BOS, L. (1970). *Symptoms of Virus Diseases in Plants*, PUDOC, Wageningen

BRAKKE, M. K. (1970). Systemic infections for the assay of plant viruses. *Annual Review Phytopathology*, 8, 61-84

CADMAN, C. H. (1963). Biology of soil borne viruses. *Annual Review Phytopathology*, 1, 143-72

DIENER, T. O. (1963). Physiology of virus-infected plants. *Annual Review Phytopathology*, 1, 197-218

DUFFUS, J. E. (1971). Role of weeds in the incidence of virus diseases. *Annual Review Phytopathology*, 9, 319-40

ESAU, K. (1967). Anatomy of plant virus infections. *Annual Review Phytopathology*, 5, 45-76

FRAZIER, N. W., FULTON, J. P., THRESH, J. M., CONVERSE, R. H., VARNEY, E. H. and HEWITT, W. B. (eds) (1970). *Virus Diseases of Small Fruits and Grapevines—a Handbook*, University of California Division of Agricultural Sciences, Berkeley

FULTON, R. W. (1966). Mechanical transmission of viruses of woody plants. *Annual Review Phytopathology*, 4, 79-102

HARRIS, K. F. (1981). Arthropod and nematode vectors of plant viruses. *Annual Review Phytopathology*, 19, 391-426

HARRISON, B. D. (1977). Ecology and control of viruses with soil inhabiting vectors. *Annual Review Phytopathology*, 15, 331-60

HOLLINGS, M. and STONE, O. M. (1964). Investigation of carnation viruses. 1. Carnation mottle. *Annals of Applied Biology*, **53**, 103–18

HOLMES, F. O. (1960). Rose mosaic cured by heat treatments. *Plant Disease Reporter*, **44**, 46–7

JANICK, J. and MOORE, J. N. (eds) (1975). *Advances in Fruit Breeding*, Purdue University Press, Indiana

KASSANIS, B. (1955). Some properties of four viruses isolated from carnation plants. *Annals of Applied Biology*, **43**, 103–13

MARKHAM, R. (1977). Landmarks in plant virology: genesis of concepts. *Annual Review Phytopathology*, **15**, 17–39

MATTHEWS, R. E. F. (1970). *Plant Virology*, Academic Press, London

McWHORTER, F. P. (1965). Plant virus inclusions. *Annual Review Phytopathology*, **3**, 287–312

MUNDRY, K. W. (1963). Plant virus-host cell relations. *Annual Review Phytopathology*, **1**, 173–96

MURANT, A. F. (1974). *Viruses affecting raspberry in Scotland*. Bulletin No. 9, 37–43, Scottish Horticultural Research Institute, Dundee

POSNETTE, A. F. and CROPLEY, R. (1956). Apple mosaic viruses, host reactions and strain interference. *Journal of Horticultural Science*, **31**, 119–33

SIEGEL, A. and ZAITLIN, M. (1964). Infection process in plant virus disease. *Annual Review Phytopathology*, **2**, 179–202

SMITH, K. M. (1972). *A Textbook of Plant Virus Diseases*, Longman, London

SMITH, K. M. (1976). Insects and other arthropods as vectors of plant viruses, chapter 17, 204–220, in *Virus-Insect Relationships*, Longman, London

SWENSON, K. G. (1968). Role of aphids in the ecology of plant viruses. *Annual Review Phytopathology*, **6**, 351–74

SYLVESTER, E. S. (1980). Circulative and propagative virus transmission by aphids. *Annual Review Entomology*, **25**, 257–86

TOMLINSON, J. A. (1964). Purification and properties of lettuce mosaic virus. *Annals of Applied Biology*, **53**, 95–102

TOMLINSON, J. A. and SMITH, B. R. (1958). Big vein disease of lettuce in Britain. *Plant Pathology*, **7**, 19–22

4 BACTERIA

4.1 THE NATURE OF BACTERIA

Bacteria are single-cellular organisms that are very variable in size and shape; often a particular species possesses more than one shape (*pleomorphic*). Together with the blue-green algae they form a group known as *prokaryotes* whereas the fungi and other plants are known as *eukaryotes*. There are three general morphological forms of prokaryote cell: *cocci*, which are ellipsoidal or spherical, *bacilli*, which are cylindrical or rod-like and *spirilla*, which are spiral or helical. Included within spirilla are the *vibrio* shape, consisting of a rudimentary helix. Some other shapes are found; of particular interest to pathologists are the Actinomycetes which have a filamentous structure analogous to that of some fungi (section 4.4). Typical features of eukaryotic cells such as cytoplasmic streaming, mitochondria, chloroplasts, Golgi apparatus, endoplasmic reticular and membrane-bound vacuoles are absent in bacteria. The cytoplasm is surrounded by a triple-layered *cytoplasmic (cell) membrane* which consists of two electron-dense layers surrounding an electron-transparent layer. This structure is typical of semi-permeable membranes found in other organisms; it does, however, contain part of the respiratory and, in some cases, photosynthetic apparatus. Noticeably, sterols are absent from this membrane. Mitosis has not been observed within the nuclei of prokaryotes. The nucleus is a diffuse area containing fibrous material but lacking a nuclear membrane. Within the nuclear area is a single circular chromosome thread of double stranded DNA; its presence has been confirmed in only a few bacteria. Since there is only one chromosome, a nuclear membrane and complex mitotic process are unnecessary. There may be an attachment between a specific point on the chromosome and the cytoplasmic membrane. The initial stage of nuclear division may involve duplication of this attachment site with subsequent progressive division of the DNA thread. The two daughter nuclei are then drawn apart by separation of these attachment points. Prokaryote ribosomes are composed of two sub-units with sedimentation constants of 30 S and 50 S (Svedberg units) which combine forming a characteristic 70 S ribosome. These two components contain roughly equal amounts of RNA and protein. They are mainly found free within the cytoplasm rather than attached to membranes as in eukaryotes. In general, prokaryote cells are characterised by a considerably higher rate of multiplication than eukaryotes. This is reflected in the number of ribosomes per unit mass. In rapidly growing bacteria, ribosomes may compose up to 40 per cent of the cell dry weight. *Storage* or *inclusion granules* within the cytoplasm are of four main types: polysaccharide (glycogen and starch), lipid (poly β hydroxybutyrate and neutral lipid), volutin (polyphosphate) and sulphur. The numbers and sizes of these granules vary with environmental conditions. The *cell wall* is a rigid structure which provides shape to the cell and protection from adverse composed of 10 per cent mucopeptide, the rest being protein, polysaccharide and lipid. In a few prokaryotes the cell wall is absent (section 4.4). a thick amorphous single-layered structure mainly composed of mucopeptide; in Gram-negative types the cell wall is a multilayered structure composed of 10 per cent mucopeptide, the rest being protein, polysaccharide and lipid. In a few prokaryotes the cell wall is absent (section 4.4). Some prokaryotes are surrounded by a gel layer (*capsule*) formed of a viscous, aqueous mixture of polysaccharide. This provides protection from invasion, for example by viruses, and prevents rapid loss or gain of water in the recurrent dehydration and hydration in environments such as soil. Additionally the capsule has an ion exchange capacity which may aid in the concentration and uptake of essential cations.

Most motile bacteria possess long thin (*circa* 20 nm diameter) extra-cellular appendages (*flagella*). These are attached at one end through the cell wall to the cytoplasmic membrane by a *terminal hook* and/or *basal body*. Flagella are composed of protein (*flagellin*). The arrangement and number of flagella is a useful character for bacterial identification and classification. Their function is to provide a means of locomotion. Some prokaryotes lack flagella but are still motile; this is accomplished by at least two means.

1. *Gliding* This requires contact by the cell with a surface and probably results from some form of contractile elements contained in the outer cell layers.
2. *Spirochaetal movement* This is typical of *Spirochaeta* spp. which have helically shaped cells with an axial filament, or filament composed of flagellin-like elements, built into the cell wall. Motion may take place by alternate expansion and contraction of this filament.

Some bacteria, particularly Gram-negative types, possess another form of extra-cellular appendage, the *pili (fimbriae)* which are composed of protein sub-units (*pilin*). These tend to be shorter, smaller and more numerous than flagella, that is, 0.5 µm long by 9 nm wide and number 1–400 per cell. They lack the characteristic wavy form of flagella. Their functions include providing a means of attachment and also act as a conjugation tube through which DNA may pass between prokaryote cells.

A comparison of the major features of prokaryote and eukaryote cells is given in table 4.1.

Table 4.1

Comparison of the main features of prokaryote and eukaryote cells

	Prokaryote	*Eukaryote*
Typical diameter	1 μm	10 μm
Number of chromosomes	1	> 1
Nuclear membrane	Absent	Present
Mitosis	Absent	Present
Number of compartments	Usually 1	> 1
Oxidative phosphorylation site	Cytoplasmic membrane	Mitochondria
Photosynthetic phosphorylation site	Cytoplasmic membrane	Chloroplasts
Vacuoles	Rare	Common
Cytoplasmic ribosome size	70 S *	80 S *
Peptidoglycan	Usually present	Absent
Flagella	1 fibril	Many fibrils plus membrane
Sexual reproduction	Rare and incomplete	Common and complete

* S = Svedberg units

Bacteria are extremely versatile in their nutritional abilities. For example, some utilise atmospheric nitrogen, whereas others require inorganic nitrogen in the form of nitrates, nitrites or ammonia and a further group utilise organic compounds such as proteins or amino acids. An interrelated cycle of these organisms is essential for the maintenance of soil nitrogen levels. Another group is able to utilise inorganic and organic forms of sulphur. Most bacteria, however, have macro and micro nutrient requirements similar to higher plants. Some possess the necessary synthetic machinery to produce their entire nutritional requirements, while others require an external supply of specific growth factors. Such traits can be used as markers for the study of bacterial genetics. As with other living organisms, bacteria require a supply of water. Their temperature requirements vary widely: some are capable of growth at below 0°C or above 60°C while surviving much higher temperatures. Oxygen is necessary for some bacteria while others have anaerobic, facultatively anaerobic or microaerophilic modes of life. Similarly pH requirements vary with species in the wide range of pH = 4–9; indeed *Thiobacillus thiooxidans* can operate at pH < 1.0. On the basis of their wide diversity of energy-yielding processes, bacteria may be described as follows.

Heterotrophs

These obtain energy from organic matter or complex organic compounds, usually as saprophytes or parasites. This may be achieved aerobically, anaerobically or by fermentation.

Autotrophs

There are two forms of autotroph.

1. *Chemoautotrophs* These obtain energy by the oxidation of inorganic substrates, usually at the expense of oxygen from a terminal electron acceptor. Adenosine triphosphate (ATP) is formed by oxidative phosphorylation and the basic difference from energy production in aerobic heterotrophs is that the electron donor is inorganic. Typical electron donors are ammonia, nitrite, hydrogen sulphide, sulphur, hydrogen, methane and ferrous iron.
2. *Photoautotrophs* These obtain energy mainly by photosynthetic phosphorylation, a process involving the excitation of chlorophyll molecules by light so that high-energy electrons are emitted thereby causing the reduction of specific compounds of low redox potential. These reduced substances are then oxidised through an electron transport system with the simultaneous production of ATP by a process similar to that of oxidative phosphorylation. The result is that the electron is returned to the chlorophyll molecule. A basic difference between this process in prokaryotes and in higher plants is that, in the former, electrons are returned to the chlorophyll by an external electron donor such as inorganic sulphur compounds, consequently no oxygen is evolved.

Prokaryotes possess a wide range of methods for obtaining energy and this is reflected in an ability to colonise a great diversity of habitats. These processes are discussed in greater detail by Stanier *et al*. (1971), Pelczar *et al*. (1977) and Wilkinson (1982).

Multiplication is achieved in many prokaryotes by an asexual process, *binary fission*, in which individual cells at first elongate, and then the cell walls begin to invaginate with a concomitant distribution of nuclear material between two emerging portions of each cell. Invagination ultimately results in the formation of transverse cell walls with nuclear material contained in each separate cell. The progeny cells separate and are quickly capable of repeating this process. The time taken for single cells to divide by binary fission, and thereby to double the population, is short and is known as the *generation time*. *Escherichia coli*, for example, divides at approximately 20-minute intervals. Multiplication of many bacterial plant pathogens occurs for short periods in both resistant and susceptible hosts. After this the population in resistant hosts declines whereas rapid multiplication continues in susceptible types.

Resting bodies are developed by some prokaryotes, particularly those that inhabit soil; these form *spores* capable of resisting adverse environmental conditions. Characteristically, a single *endospore* is formed within a vegetative cell which on germination re-establishes a single vegetative cell. Two other resting structures form less frequently: *cysts*, which are intermediate in structure and resistance properties between endospores and vegetative cells and *exospores (conidia)* which form by fragmentation of the aerial ends of multinucleate filaments of the Actinomycetes (section 4.4).

Sexual recombination in the prokaryotes is less well developed than in eukaryotes. It takes place sporadically and does not involve 'true' fusion of male and female gametes resulting in the formation of a diploid zygote. There are three mechanisms whereby DNA can pass from donor to recipient cells in prokaryotes.

Conjugation

This process requires cell-to-cell contact (Lederberg and Tatum, 1946) and there is unidirectional transfer of genetic material indicating some level of sexual differentiation between conjugating cells. Transfer of DNA from the donor or 'male' cell to the recipient or 'female' cell involves special pili (F, fertility, pili or sex-pili) which form a bridge between the cells (Hayes, 1968). The DNA passes through a channel in the centre of a pilus resulting in an orderly entry of genes according to their arrangement on the chromosome. Theoretically the whole chromosome can be transferred but in practice the process is rarely completed.

Transformation

Some bacteria, when grown in the presence of dead cells, culture filtrates or cell extracts (DNA) from a closely related strain, will acquire and subsequently transmit characteristics of that related strain. Direct transfer of DNA material between strains has been established in some plant pathogenic bacteria (Corey and Starr, 1957a, b).

Transduction

This requires a bacteriophage (a virus parasitic to bacteria) to inject viral nucleic acid into a bacterial cell where it replicates (section 3.3) forming further bacteriophage proteins. In some instances, viral genomic material becomes integrated with that of the bacterium. This hybrid material is released and invades another bacterium. Consequently, transduction is the transfer of a portion of bacterial DNA from one cell to another by a bacteriophage serving as a vector (Loutit, 1958; Holloway and Monk, 1959).

Plasmids

These are forms of extra-chromosomal DNA capable of existing within a bacterial cell in an autonomously reproducing state and dividing in phase with normal cell division; they behave like secondary chromosomes. They are composed of genes controlling properties that are additional to those required for normal cell growth such as resistance to antimicrobial agents or pathogenicity to particular hosts. The latter is especially important in relation to the crown gall-inducing pathogen *Agrobacterium tumefaciens* (section 4.2). Resistance plasmids pose a considerable problem to the use of chemotherapy as a means of controlling bacteria since their transfer to previously sensitive bacteria may result in their acquiring resistance to antibiotics. Further information concerning plasmids is given by Broda (1979), Day (1982) and Hepburn (1982).

4.2 EFFECTS OF BACTERIAL INFECTION ON HOST PLANTS

Bacteria are unable to penetrate directly through the plant cuticle; they invade via aerial wounds, stomata, other orifices, or wounds formed by secondary roots as they emerge from parent roots (Kelman and Sequeira, 1962). Unlike some other types of plant pathogen, bacteria do not make immediate direct contact with host protoplasm following invasion. They are mainly restricted to intercellular spaces or vascular elements where initial multiplication takes place without intracellular contact (Nelson and Dickey, 1970). Bacteria may be divided into the following three groups based on their primary effects on a host.

Soft rot bacteria

These cause cell wall maceration because of enzymic action.

Pre-necrotic and necrotic bacteria

These cause yellowing, wilting, blights, aerial spotting and cankers usually associated with altered cell physiology and metabolism; there are many gradations and forms of pathogenesis within this group.

Hypertrophic bacteria

These cause galls, fasciations and scabs following altered growth-control mechanisms and the induction of autonomous host growth (Kelman, 1969).

Bacteria that are plant pathogens are usually facultative parasites, eventually killing the host tissues. Once this is accomplished, their competitive saprophytic abilities are limited in comparison with other microorganisms. Non-pathogenic bacteria predominate over pathogenic ones in the flora around and on higher plants. Bacteria lack, for example, the ability found in many fungi for wind dispersal or the efficient vector relationships of some viruses. Judged by the criterion of number of plant-pathogenic compared with non-pathogenic bacteria, they do not appear to be successful plant pathogens. There are, however, individual species such as *Pseudomonas solanacearum*, a cause of bacterial wilt in a range of tropical crops, that are severe and aggressive pathogens. Attributes that aid bacteria as plant pathogens include the following.

1. The possession of resistant cell walls and a capacity to produce slime (ooze) and spores, both of which enable them to survive adverse environmental conditions.
2. A rapid and simple mode of multiplication, which permits large populations to develop quickly.
3. Simple growth requirements.
4. Possession of enzyme-producing and possibly toxin-producing systems, which can modify the host and thereby place the pathogen at an advantage.

As with other plant pathogens, the general course of a bacterial host–parasite relationship may be viewed in phases.

1. *Affinity*—compatibility between host and parasite.
2. *Aggressiveness*—a capacity to invade a host, that is, the ability to enter and inhabit, overcoming resistance mechanisms and multiplying at the expense of a host.
3. *Pathogenicity*—an ability to evoke disease symptoms in the host (Gäumann, 1951).

At least some plant-pathogenic bacteria release into a host one or more of a broad range of physiologically active compounds including pectolytic, proteolytic and, occasionally, cellulolytic enzymes, phosphatidases, growth regulators (indoleacetic acid, ethylene and cytokinins), peptidoglycans, polysaccharides and low molecular mass toxins. The potential roles that these compounds might play in the initial stages of pathogenesis have been examined by numerous workers. Only in a few instances is there a clear cause-and-effect relationship (Kelman and Sequeira, 1972). There is scant evidence that any of these products are primary determinants of host specificity, nor have they been shown conclusively to play a primary role in the induction of specific host defence reactions. Some of these materials accumulate in the host at the time of symptom expression and it may be logical to ascribe to them a role such as the following.

Wilt toxins

Certain polysaccharides are formed by phytopathogenic bacteria *in vitro* that possess wilt-inducing properties. Generally these substances are far less host-specific than the micro-organisms that form them.

Necrotising toxins

In wild fire disease of tobacco caused by *Pseudomonas tabaci*, for example, a characteristic symptom on leaves is the formation of angular, brown, necrotic spots surrounded by broad chlorotic haloes; these are free from bacteria. Cell-free filtrates of *P. tabaci* will induce similar chlorotic haloes if injected into tobacco leaves. This toxin is characterised as a lactone of α-lactyl amino-β-hydroxy-ϵ-amino pimelic acid. Substances with similar properties have been implicated in halo blight of *Phaseolus* bean (*Pseudomonas phaseolicola*—section 4.5) and stone-fruit necroses.

Hypertrophy-inducing factors

An example of these is the tumourous growth associated with *Agrobacterium tumefaciens* (crown gall). It is suggested that large extra-chromosomal DNA molecules (*plasmids*) confer tumourigenic and virulence properties on *A. tumefaciens*. The surface components of *A. tumefaciens* alter when they carry plasmids and these changes permit the bacteria to attach to specific target sites in wounded host tissue, subsequently allowing a transfer of further unidentified tumourigenic substances. It is thought that tumours result from host cells producing at least two uncommon amino acids, octopine and nopaline (Kado, 1976; Dixon, 1984).

Rot-inducing enzymes

The soft-rot symptoms associated with some bacteria result from enzymic decomposition of substances within the cell wall. Prominent enzymes causing these effects are as follows.

1. *Pectin esterase (PE)*—this hydrolyses the methyl-ester groups of pectin ultimately yielding demethoxylated pectic acid.
2. *Polygalacturonase (PG)*—of which there are several types implicated in hydrolysis of α-1,4 glycosidic linkages of pectin and pectic acid.
3. *Cellulase (Cx)*—though of lesser significance in pathogenesis caused by bacteria than that caused by fungi, these enzymes have been implicated in changes to the constitution of host cell-wall cellulose and reduction of its polymerisation for instance when tomato and tobacco stems are infected by *Pseudomonas solanacearum* (Kelman and Cowling, 1965).

The virulence of bacteria causing soft rots has been correlated with their production of enzymes capable of degrading cell-wall components (Bateman and Millar, 1966). The restoration of virulence in a mutant strain of *Erwinia carotovora* has been associated with an ability to produce pectic, cellulolytic and phosphatidase systems (Beraha and Garber, 1971).

4.3 SURVIVAL AND SPREAD

The success of any pathogen relies in part on the amount of inoculum that can be produced over a period. Owing to their rapid generation rate, a small quantity of surviving bacteria may quickly produce an epidemic. Plant-pathogenic bacteria do not form resting spores comparable with those of fungi and must in consequence remain dormant in association with seeds, perennial host or plant parts, insects, epiphytes, plant residues, soil and other non-host materials (Schuster and Coyne, 1974).

Seed-borne inoculum

At least 95 species of plant-pathogenic bacteria (approximately 50 per cent of the total) are seed-borne. They can survive for prolonged periods in seed; for example *Corynebacterium flaccumfaciens* survives with *Phaseolus* bean seed for 5–24 years. The mode of seed germination can affect the efficacy of subsequent spread of seed-borne inoculum. Plants with epigeal germination (where the cotyledons form the first foliage leaves), for example, beans, beet, cabbage, carrot, garden stocks, lettuce, pepper and tomato, favour transmission from seed-borne sources. Examples of bacteria that spread in this way are *Xanthomonas campestris* on cabbage, *X. carotae* on carrot, *X. incanae* on stocks, *Pseudomonas phaseolicola* on bean (halo blight—section 4.5) and *Corynebacterium michiganense* on tomato. Conversely, this mode of germination severely limits the ability of a pathogen to invade subterranean host organs. The success of seed-borne bacteria depends on location on the seed, anatomical structure of the seed and survival capacity of the bacteria.

Inoculum associated with plant residues

Even when 'pathogen-free' seed is used, regular epidemics can develop if crop rotation is not practised. For example, bean-crop trash is an important overwintering source of *Pseudomonas phaseolicola* (halo blight—section 4.5).

Association with perennial hosts

Only a few cankers induced by *Erwinia amylovora* (fire-blight) contain viable bacteria but these are sufficient to initiate regular spring epidemics in the USA.

Epiphytes as inoculum sources

Epiphytic bacteria (those living on plant surfaces) can be a primary source of inoculum. *Pseudomonas syringae* (canker and shoot blight) may be recovered throughout the year from leaf surfaces of *Vicia villosa* (hairy vetch) and can be associated with outbreaks of bean brown-spot disease (Hagedorn *et al*., 1972). Similarly, phylloplane growth of *P. mors-prunorum*

(*P. syringae*) may be a primary source of inoculum for infections of stone-fruit tree branches in autumn (Crosse, 1968).

Soil-borne bacteria

These may be divided into the following three groups in relation to soil (Buddenhagen, 1965).

1. *Pathogens whose populations develop almost exclusively in host tissue.* A soil phase, if it exists, declines rapidly and does not usually contribute to the propagation of disease from season to season. Many plant pathogens fit into this group; examples are *Erwinia amylovora* (fire-blight), *Corynebacterium sepedonicum* (potato ring rot) and most *Xanthomonas* spp.
2. *Pathogens that build up in a host but whose populations decline only slowly in soil.* Dependent on cropping practice, there could be gradual increase, decrease or eventual extinction of a soil-borne population. Long-term presence in soil is host-mediated and hence these are not true soil inhabitants; examples are *Agrobacterium tumefaciens* (crown gall), *Pseudomonas solanacearum* (wilt of Solanaceae) and *Erwinia carotovora* (soft rot).
3. *Pathogens whose populations are largely produced in the soil or rhizoplane.* A relationship to plant disease may be erratic and transitory as for example with some green-fluorescent *Pseudomonas* spp. that cause soft rots.

Other non-host sources of inoculum

Bacteria such as *Corynebacterium sepedonicum* (potato ring rot) persist on harvesting and grading machinery, sacks and storage bins. Eradication is difficult because the bacterial cells are highly resistant to heat and desiccation.

The spread of bacteria can be considered at two levels: transmission of primary inoculum to new crops and distribution of secondary inoculum within a crop. Bacteria, unlike fungi, possess no adaptive structures for launching cells into air currents and cannot penetrate intact plant surfaces. Spread is dependent on agencies such as propagating material, soil, rain, wind, insects and husbandry implements. Primary spread takes place mainly via infected propagating material and soil; the other agencies are largely responsible for secondary spread. Rain permits the mobilisation of inoculum which is normally formed within a host and reaches the plant surface as slime or ooze. It will also effect the entry of bacteria through wounds and natural openings, possibly causing waterlogging of intercellular spaces whereby bacteria can invade in a continuous film of water between the interior and exterior of a leaf. Irrigation water and cultivation implements provide means for a movement of bacteria within crops.

Spread via insect vectors has not been proved in many cases, but it is likely over short distances. A few bacteria are capable of survival within an insect body. *Erwinia tracheiphila* (cucurbit wilt) is completely dependent on cucumber beetles for survival between seasons and subsequent spread. Primary inoculum in spring enters via feeding-punctures caused by over-wintered beetles. Transmission is probably through food regurgitation (section 3.2). Bacterial strands less than 1 mm long and composed of cells bound together with ooze produced by *Erwinia amylovora* (fire-blight) may be wafted short distances in wind currents, transported in the dry state over longer distances or carried upwards by wind currents into nearby clouds where they are dispersed in moisture and subsequently fall in raindrops at a distant location.

Bacterial slime or ooze, a mechanism for added protection, is frequently found in infected seed, cankers and living or dead plant parts. *Corynebacterium flaccumfaciens* (wilt of beans), for instance, can survive for up to 5 years in dried ooze. Similarly *Erwinia amylovora* (fire-blight) survives for long periods in dry ooze without which this organism is susceptible to desiccation. Dry conditions can, however, favour the survival of those bacterial pathogens that overwinter in plant residues. In these conditions, plant residues are not degraded and thereby protect the bacteria from predatory soil fauna such as protozoa and nematodes.

Bacteria would not survive without means of distribution and survival over winter in temperate zones and dry periods in tropical areas. These are points at which control measures are likely to be most effective, either directly or when primary inoculum is giving rise to new epidemics. Controls include eradication, pathogen-free planting material, resistant cultivars and modified cultural practices.

4.4 BACTERIAL CLASSIFICATION

Characteristics used in the identification and classification of bacteria largely rest on chemical and cultural properties of populations rather than on morphological attributes of individuals as employed with higher plants. Examples of some of these characters are shown in table 4.2. In addition, since bacteria reproduce predominantly by an asexual process, it is difficult to interpret them in terms of species as defined for higher organisms where this term is largely based on the understanding of sexual reproduction. Consequently, species of bacteria and those of higher organisms where this term is largely based on an understanding of sexual is difficult to arrange bacteria in a hierarchical system that reflects phyllogenetic relationships. Various investigators, therefore, have made considerable rearrangements in nomenclature even within the small group of plant pathogens. There are perhaps only about 200 species of plant pathogens among the much larger total group of bacteria. It was not until the 1880s,

Table 4.2

Some characters used to identify bacteria

1. *Colony characters in vitro*
   - 1.1 Colony colour, abundance of growth (scanty, moderate, abundant), production of odours (putrid, fruity, aromatic)
   - 1.2 Colony size (small, large)
   - 1.3 Shape of colony margins (entire, round, irregular)
   - 1.4 Formation of thread-like growths into agar
   - 1.5 Elevation of colonies (thin, flat, thick, raised)
   - 1.6 Optical qualities (opaque, translucent, opalescent)
   - 1.7 Colony constituency (butyrous, viscous, stringy, dry, brittle)
   - 1.8 Formation of a skin (*pellicle*) or sediment in liquid culture
   - 1.9 Liquefaction of gelatin

2. *Staining characters*
   - 2.1 Reaction to specific reagents, for example, Gram's

3. *Biochemical effects*
   - 3.1 Production of gases (hydrogen sulphide, ammonia)
   - 3.2 Synthesis and degradation of organic compounds
   - 3.3 pH changes in milk/litmus mixture
   - 3.4 Sensitivity to antibiotics
   - 3.5 Serological reactions
   - 3.6 Bacteriophage reactions
   - 3.7 Estimation of DNA base ratios

Note: For greater detail see Conn and Pelczar (1957), Collins (1964), Harrigan and McCance (1966), and Skerman (1969).

following work on fire-blight disease of apple and pear in the USA and yellows disease of hyacinth in Holland, that bacteria began to be accepted as plant pathogens (early work in this field was reviewed by Jones, 1939). Early in this century, plant-pathogenic types were classified separately from other bacteria solely on the basis that they were plant pathogens. Work by Burkholder (1930), Conn (1942) and Dowson (1939, 1942, 1943) brought them within the ambit of general bacteriology (reviewed by Starr, 1959). Since then there have been various attempts to combine and divide genera on a range of criteria.

A general outline of the taxonomic positions currently accepted for plant pathogenic bacteria is shown in table 4.3. The species position is confused because many names have been based on limited identification of isolates taken from new hosts (the new host ≡ new pathogen species concept). This has led to a plethora of bacterial species names of tenuous validity. It is now considered that there are 'relatively few true species of phytopathogenic bacteria (that) exist in many specialised forms character-

Table 4.3

Outline groupings of plant-pathogenic bacteria (after Buchunan and Gibbons, 1974)

1. *Gram-negative, aerobic rods and cocci*		
	Family	Pseudomonadaceae
	Genera	*Pseudomonas*
		Xanthomonas
	Family	Rhizobiaceae
	Genus	*Agrobacterium*
2. *Gram-negative, facultatively anaerobic rods*		
	Family	Enterobacteriaceae
	Genus	*Erwinia*
3. *Actinomycetes and related organisms*		
	Coryneform group	
	Genus	*Corynebacterium*
	Family	Streptomycetaceae
	Genus	*Streptomyces*
4. *Rickettsia**		
	Family	Rickettsiaceae
	Genus	*Rickettsia*
5. *Mycoplasma**		
	Family	Mycoplasmataceae
	Genus	*Mycoplasma*
	Genus	*Spiroplasma* (uncertain affiliation)
	Mycoplasma-like bodies in plants	

* Recognised as potential plant pathogens since 1965.

ised by a quite elastic adaptability to particular host plants' (Starr, 1959). As with fungi and viruses, nomenclature is further confused by the presence of intraspecies biotypes and strains. This was first noted by Ark (1937) in *Erwinia amylovora* (fire-blight). Studies of ten single-cell isolations of *E. amylovora* from eight host types grown in six localities showed appreciable differences in the length of bacterial cells, rate of growth, size and form of colonies on nutrient media, physiological characters and virulence. Similar results were obtained by Decker and Becker (1939) who demonstrated two 'pathogenic' races of *Streptomyces scabies* (potato common scab) that were similar but not identical in cultural characteristics and to which some potato seedlings were resistant to one race and not to another. Phytopathogenic bacteria have now been shown to exhibit considerable plasticity of host range (Dye, 1958; Dye *et al.*, 1964), and there are authenticated cases of acquisition and loss of virulence based on reproducible and confirmed experiments (Stolp, 1961).

Bacterial genera that contain plant-pathogenic types and can be defined with reasonable clarity are as follows.

Agrobacterium

These comprise small, motile, short rods with one to four peritrichous flagella. They are ordinarily Gram-negative and do not produce visible gas or detectable acid in culture media; gelatin is liquified slowly or not at all. They do not fix free nitrogen but utilise inorganic forms of nitrogen; optimum temperatures for growth are 25–30°C. Species are found in soil or plant roots, in hypertrophies (galls) on roots and stems of plants.

Corynebacterium

These comprise slender, straight or slightly curved rods with irregularly stained segments or granules, often with pointed or club-shaped swellings at the ends. They are generally non-motile and Gram-positive. They are widely distributed in nature, causing cankers and wilts in crop plants.

Erwinia

These comprise motile rods that normally do not require organic nitrogen for growth. They produce acids without gas from a variety of sugars, and many possess pectolytic systems. They invade plant tissues to produce dry necroses, galls, wilts and soft rots. *Erwinia* spp. belong to the same family as *Escherichia coli* which has been studied extensively as a model biological system.

Pseudomonas

These comprise motile, straight rods with polar flagella. They are Gram-negative, many species producing a greenish fluorescent water-soluble pigment. Various species are found in soil and water. The plant-pathogenic types cause leaf spots, wilts and cankers.

Xanthomonas

These comprise small rods that are motile with a single polar flagellum. They form abundant slimy yellow growth in certain culture media. Plant-pathogenic types cause extensive necrosis of the host.

Streptomyces

Within the family Streptomycetaceae are species that exhibit some characteristics akin to fungi although with bacterial dimensions and in consequence are termed *mould-like bacteria*. These characters include the production of branched filaments which form spores similar to those developed by some fungi. For example, *Streptomyces scabies* (potato common scab) forms vegetative filaments that do not fragment easily and bear spores at the tips of aerial hyphae. This genus is Gram-positive, aerobic and primarily soil-inhabiting.

Mycoplasma, spiroplasma and rickettsia-like organisms

Before 1965 the causal agents of yellows diseases were thought to be viruses. It is now apparent that members of these bacterial taxons are likely to be causal agents of yellows and a range of other diseases (Doi *et al*., 1967; Ishie *et al*., 1967). Strict proofs of pathogenicity for specific diseases are lacking, however, because of difficulties in cultivating these organisms *in vitro*. Consequently the terms *Mycoplasma-like organisms (MLO)* (which may include *spiroplasma*) and *rickettsia-like organisms (RLO)* are used. These organisms are possibly the simplest life forms capable of independent growth and metabolism. MLOs have been associated with 200 plant species. Species of *Mycoplasma* may have several forms (*pleomorphism*) with coccoid, filamentous and bulbous types in one isolate. Some are capable of passing through a bacteriological filter (hence the confusion with viruses), the smallest ranging in size from 125 to 240 nm. Characteristically, minute colonies form on nutrient agar and in living cells. Two modes of reproduction have been identified: segmentation and formation of elementary units surrounded by a membrane. MLOs have a triple-layered membrane similar to that of true *Mycoplasma* with a central cell structure containing fibrillar material which is presumably DNA; cell vacuoles are frequently seen particularly in the globular types. *Spiroplasma* are more easily cultured *in vitro* than *Mycoplasma*. Even so, pathogenicity has been proved only for *S. citri*, the cause of Stubborn and Little Leaf Disease of Citrus. *Spiroplasma* have helical bodies 3–25 μm long and 100–200 nm diameter which may alter in culture to rounded forms. Three-dimensional structures occur on the surface of *Spiroplasma* which resemble those of true *Mycoplasma*.

RLOs are mostly found in xylem, phloem and meristematic tissues of diseased plants. They closely resemble true *Rickettsia*, with rod-shaped cells 0.2–0.5 μm wide and 1.0–3.0 μm long, and possess a well-defined cell wall and plasma membrane. Some RLOs are pleomorphic displaying spherical or filamentous shapes. Internally they possess electron-dense areas and one or more areas of low density which contain DNA material similar to that of other bacteria. Ribosome-like structures have been reported.

MLOs are usually confined to sieve tubes but are occasionally found in phloem parenchyma and companion cells but infrequently in xylem, mesophyll cells or root pericycles. They occur in the cytoplasm of mature cells and in meristematic tissue at shoot apices. Both MLOs and RLOs can reach densities of 100 particles per host cell. Transmission from plant to plant requires an insect vector, for which 60 spp. of leaf hopper, some aphids and psyllids have been implicated. Probably there is also transmission during vegetative propagation and by mechanical means. MLOs in particular require a long incubation period in leaf hoppers while the acquisition phase varies from 3 hours to 5 days. Diseases attributed to these organisms are mainly of economic significance in warm climates favouring the vectors. MLOs disrupt the normal host growth regulator concentrations leading to formation of adventitious buds, altered apical dominance, proliferation of axillary shoots and witches' broom symptoms. Four types of disease syndrome may be identified.

1. *Aster yellows*—typified by elongation of internodes and foliar yellowing.
2. *Stolbur*—typified by apical dwarfing, stunting, leaf-roll, epinasty, wilting and virescence (development of green colouration in tissues where it is normally absent, for example, in the perianth).
3. *Witches' broom*—typified by a proliferation of axillary shoots.
4. *Decline*—typified by host degeneration.

Several symptom types may be caused by the same MLO.

Diseases caused by RLOs may be grouped as follows.

1. *Xylem-limited types* These resemble vascular wilt diseases where there is disrupted water conduction, decline in vigour, marginal necrosis or scorching of foliage, decreased productivity, small fruit and eventual host death. Organisms causing these symptoms differ from other vascular invading bacteria in that RLOs are rarely found outside xylem tracheids and vessels, they do not form lysigenous cavities and they are introduced directly into the xylem by a vector. Diseases of horticultural crops associated with xylem-limited RLOs include almond leaf scorch, phony disease of peach, Pierce's disease of grapevine (the most intensively studied) and plum leaf scald.
2. *Phloem-limited types* These are characterised by the yellows type of syndrome in which symptoms include stunting, yellowing of young leaves, leaf-curling and twisting, uneven expansion of leaves, virescence of flower petals and premature death of the entire plant. Diseases of horticultural crops associated with phloem-limited RLOs include citrus greening, little leaf of *Sida cordifolia* and potato leaflet stunt.

A third group of diseases that cannot be ascribed specifically to either xylem-limited or phloem-limited RLOs includes apple proliferation*, carrot proliferation*, infectious necrosis of grapevine and yellows of grapevine (* = may also be caused by MLOs).

Control methods are similar to those employed for viruses; they include heat therapy, meristem tip culture, selection of symptomless mother stocks, control of vectors by organophosphorus and oximecarbamate insecticides and control of weed hosts. In isolated cases chemotherapy may be effective; tetracyline gives remission of mulberry dwarf symptoms. Although antibiotics of the penicillin type are not effective against true *Mycoplasma* and MLOs they do have some effect on RLOs.

L-phase bacterial variants

Several species of bacteria go through a transition to form very small bodies termed L-phase bacteria (L after the Lister Institute). The term L-phase designates an independent growth variant of a bacterium that lacks a rigid cell wall but retains a potential to revert to the parental bacterium (Hijmans *et al.*, 1969). In morphology they appear to be somewhat similar to *Mycoplasma*. Some L-phase bacteria arise apparently by colony mutation whereas others result from the action of inhibiting agents such as antibiotics like penicillin or enzymes, for example, lysozyme.

4.5 AN EXAMPLE OF A PLANT-PATHOGENIC BACTERIUM–*PSEUDOMONAS PHASEOLICOLA*–THE CAUSE OF HALO BLIGHT IN *PHASEOLUS* BEAN

Symptoms

This pathogen is widely distributed in all areas of the world where species of *Phaseolus* bean are produced and causes considerable crop losses. The seed-borne nature of this bacterium permits such widespread dispersal. Halo blight disease is often referred to as 'grease spot disease' because of the greasy water-soaked lesions found on infected organs, which result from tissue degradation caused by enzymes. In the early stages of infection host plants have a yellowish discolouration to the leaves but, as the pathogen becomes systemically established, symptoms may vary. Infection is often initiated in the minor veins progressing to the main vein. Invasion is via stomata and hydathodes, typifying bacterial entry through orifices. Reddish colours develop in the interveinal areas; if infection starts in the petiole, this and adjacent veins first take on a reddish appearance. Water-soaked spots develop on the undersurface of leaflets which gradually enlarge and coalesce. Later a halo-like zone of greenish yellow tissue develops outside the water-soaked region. Single haloes may be 2–3 cm in diameter. On young plants, stem lesions start as water-soaked spots which enlarge and appear as reddish streaks expanding along the length of the stem. The stem epidermis and cortex may split and bacterial slime is exuded to be dispersed by water splash. Bacterial slime generated by *P. phaseolicola* is coloured light cream to silver.

Where plants develop from infected seed a primary symptom is stem-girdling or 'joint rot'. Symptoms start at the primary node above the cotyledon as a small water-soaked amber-coloured lesion which girdles the stem. Girdling is usually complete when the bean pods are semi-mature and the increasing weight of swelling pods causes the plant to break at the weakened node. Pod infections are often zonate have a small brown-to-red region of tissue surrounding the lesion. Lesions are common on pod sutures leading to invasion of the vascular system and infecting seed via the funiculus. Wilting is common in plants grown from infected seed; leaflets wilt from the pulvinus in daytime but recover their turgor at night. If young pods are infected, the seed may rot and fail to mature. If penetration is through the funiculus alone only the hilum may be discoloured which is difficult to detect on dark-coated cultivars. The plumules of seedlings grown from infected seed may be destroyed. This condition is known as 'snakeshead' and can lead to a proliferation of lateral shoots but the plants are very dwarfed.

Biology

Pseudomonas phaseolicola is an aerobic, Gram-negative, rod-shaped bacterium that occurs singly or in pairs and is motile possessing from one to six polar or bi-polar flagella. Its host range includes: *Phaseolus vulgaris*, *P. coccineus*, *P. lunatus* var. *macrocarpus*, *P. multiflorus*, *P. atropurpureus*, *Pueraria thunbergiana* and *Glycine max* (the last-mentioned on artificial inoculation). The conditions required for epidemic development are outlined in figure 4.1.

Invasion takes place through the stomata and other natural orifices or at sites of mechanical injury. Once inside the host, the whole vascular system becomes invaded and the host is systemically infected. Bacteria migrate through the intercellular spaces of parenchyma tissue, dissolving the middle lamella slightly in advance; affected cells collapse and are invaded. This tissue breakdown leads to the development of lysigenous cavities. Large bacterial masses form in the xylem cells as a result of multiplication and the cell walls either rupture or are dissolved enzymically and the pathogen passes into the parenchyma tissues again.

Control

Host resistance is used as a main avenue for the control of this pathogen. Unfortunately several physiological races of *P. phaseolicola* have evolved typifying the variability of bacterial populations. Races 1 and 2 are well documented and sources of host resistance identified but further races have appeared in France, Romania and Zimbabwe. The most important source of resistance to *P. phaseolicola* races 1 and 2 is PI 150 414 which is used in breeding programmes in the Netherlands, UK and USA. Resistance to further races may be available in OSU 10183 which has a complex multispecific parentage.

The value of pathogen-free seed in combating this pathogen cannot be over-emphasised (section 8.2.2). In addition to systemic invasion of the

Infected seed

↓

Systemically infected seedlings leading to the 'snakeshead' condition (primary infection)

↓

Joint rot and splitting of stems, reddish foliar discolouration with greasy water-soaked lesions and pod infection; maximum symptom expression at 16–20°C

↓

Exudation of bacterial slime on plant surfaces

↓

Spread by wind and rain splash up to 25 m and reinvasion through stomata and sites of mechanical injury (secondary infection)

↓

Infected plants with spread of *P. phaseolicola* through the vascular system to infect healthy foliage together with invasion of pods and seeds (→ back to Infected seed)

Figure 4.1 Infection cycle of *Pseudomonas phaseolicola*—the cause of halo blight in *Phaseolus* bean

embryo, *P. phaseolicola* is found in cracks of the seed coat caused during threshing. Such infection is difficult to eradicate by means of surface sterilisation. A policy of seed production in arid sections of the inter-montane south-western and Pacific coastal states of the USA has ensured freedom from *P. phaseolicola* in seed used for first-year crop production in areas where the pathogen is *endemic* (permanently established). Roguing bean-seed crops for infected plants has been a successful means of obtaining pathogen-free seed in New Zealand.

Chemical control is limited, as with other bacterial plant pathogens, by lack of suitable materials (section 6.1). Traditional materials such as Bordeaux mixture, copper oxychloride and copper sulphate will give some measure of control for limited periods. Recently the antibiotic kasugamycin, used as a seed treatment, has proved capable of causing significant reductions in the numbers of primary foci in the field. Use of this material is acceptable since it is not employed medically. There may, however, be a danger that antibiotic resistant strains of *P. phaseolicola* could evolve if it were used alone and extensively.

REFERENCES

ARK, P. A. (1937). Variability in the fireblight organism, *Erwinia amylovora. Phytopathology*, 27, 1-28

BATEMAN, D.F. and MILLAR, R.L. (1966). Pectic enzymes and tissue degradation. *Annual Review Phytopathology*, 4, 119-46

BERAHA, L. and GARBER, E. D. (1971). Avirulence and extracellular enzymes of *Erwinia carotovora. Phytopathologische Zeitschrift*, 70, 335-44

BRODA, P. (1979). *Plasmids*, Freeman, London

BUCHANAN, R. E. and GIBBONS, N. E. (eds) (1974). *Bergey's Manual of Determinative Bacteriology*, Williams and Wilkens, Baltimore, 8th edn.

BUDDENHAGEN, I. W. (1965). The relation of plant pathogenic bacteria to the soil. 269-84, in *Ecology of Soil Borne Plant Pathogens* (eds K. F. Baker and W. C. Snyder), John Murray, London

BURKHOLDER, W. H. (1930). The genus *Phytomonas. Phytopathology*, 20, 1-23.

COLLINS, C. H. (1964). *Microbiological Methods*, Butterworths, London

CONN, H. J. (1942). Validity of the genus *Alcaligenes. Journal of Bacteriology*, 44, 353-60

CONN, H. J. and PELCZAR, M. J. (eds) (1957). *Manual of Microbiological Methods*, McGraw-Hill, New York

COREY, R. R. and STARR, M. P. (1957a). Genetic transformation of colony type in *Xanthomonas phaseoli. Journal Bacteriology*, 74, 141-45

COREY, R. R. and STARR, M. P. (1957b). Genetic transformation of streptomycin resistance in *Xanthomonas phaseoli. Journal Bacteriology*, 74, 146-50

CROSSE, J. E. (1968). Plant pathogenic bacteria in soil. 552-72, in *The Ecology of Soil Bacteria* (eds T. R. G. Gray and D. Parkinson), University Press, Liverpool

DAY, M. J. (1982). *Plasmids*, Edward Arnold, London

DECKER, P. and BECKER, H. (1939). Pathogenic races of *Actinomyces scabies* in relation to scab resistance. *Phytopathology*, 29, 204-8

DIXON, G. R. (1984). Galls caused by pathogens and related agents, in *Plant Diseases: Infection, Damage and Loss* (eds R. K. S. Wood and J. G. Jellis), Blackwells, Oxford

DOI, Y., TERANAKA, M., YORA, K. and ASUYAMA, H. (1967). Mycoplasma or PLT group-like micro-organisms found in the phloem elements of plants infected with mulberry dwarf, potato witches'-broom, aster yellows or *Paulownia* witches'-broom. *Phytopathology Society Japan Annals*, 33, 259-66

DOWSON, W. J. (1939). On the systematic position and generic names of the gram-negative bacterial plant pathogens. *Zentralblatt Bakteriologie Parasitenkunde Infektionskrankheiten und Hygiene*, Abt. II, 100, 177-93

DOWSON, W. J. (1942). On the generic name of the gram-positive bacterial plant pathogens. *Transactions British Mycological Society*, 25, 311-14

DOWSON, W. J. (1943). On the generic names *Pseudomonas*, *Xanthomonas* and *Bacterium* for certain bacterial plant pathogens. *Transactions British Mycological Society*, 26, 4-14

DYE, D. W. (1958). Host specificity in *Xanthomonas. Nature, London*, 182, 1813-14

DYE, D. W., STARR, M. P. and STOLP, H. (1964). Taxonomic clarification of *Xanthomonas vesicatoria* based upon host specificity, bacteriophage sensitivity and cultural characteristics. *Phytopathologische Zeitschrift*, 51, 394-407

GÄUMANN, E. (1951). *Pflanzliche Infektionslehre*, Birkhauser, Basel

HAGEDORN, D. J., RAND, R. E. and ERCOLANI, G. L. (1972). Survival of *Pseudomonas syringae* on hairy vetch in relation to epidemiology of bacterial brown spot of bean. *Phytopathology*, 62, 762

HARRIGAN, W. F. and McCANCE, M. E. (1966). *Laboratory Methods in Microbiology*, Academic Press, London

HAYES, W. (1968). *The Genetics of Bacteria and their Viruses*, Wiley, New York

HEPBURN, A. G. (1982). The biology of crown gall—a plant tumour induced by *Agrobacterium tumefaciens*, 101-113, in *Bacteria and Plants* (eds M. Rhodes-Roberts and F. A. Skinner), Academic Press, London

HIJMANS, W., VAN BOVEN, C. A. P. and CLASENER, H. A. L. (1969). Fundamental biology of the L-phase bacteria. 67-143, in *The Mycoplasmatales and the L-phase of bacteria* (ed. L. Hayflick), Appleton-Century-Crofts, New York

HOLLOWAY, B.W. and MONK, M. (1959). Transduction in *Pseudomonas aeruginosa. Nature*, London, 184, 1426-27

ISHIE, T., DOI, Y., YORA, K. and ASUYAMA, H. (1967). Suppressive effects of antibiotics of the tetracycline group on symptom development of mulberry dwarf disease. *Phytopathology Society Japan Annals*, 33, 267-75.

JONES, L. R. (1939). Biographical memoir of Erwin Frink Smith, 1854-1927, *National Academy Sciences, Washington, Biographical Memoir*, 21, 1-71

KADO, C. I. (1976). The tumour-inducing substance of *Agrobacterium tumefaciens. Annual Review Phytopathology*, 14, 265-308

KELMAN, A. (1969). How bacteria damage crops. 128-54, in *How Crops Grow—A Century Later* (ed. P. R. Day). Bulletin 708, Connecticut Agricultural Experimental Station

KELMAN, A. and COWLING, E. B. (1965). Cellulolytic enzymes of *Pseudomonas solanacearum* in relation to pathogenesis. *Phytopathology*, 44, 693-95

KELMAN, A. and SEQUEIRA, L. (1962). Root infection in relation to spread of *Pseudomonas solanacearum. Phytopathology*, 52, 16 (abstract)

KELMAN, A. and SEQUEIRA, L. (1972). Resistance in plants to bacteria. *Proceedings Royal Society, London*, B181, 247-66

LEDERBERG, J. and TATUM, E. L. (1946). Novel genotypes in mixed cultures of biochemical mutants of bacteria. *Cold Spring Harbor Symposium of Quantitative Biology*, 11, 113-14

LOUTIT, J. S. (1958). A transduction like process within a single strain of *Pseudomonas aeruginosa. Journal General Microbiology*, 18, 315-19

NELSON, P. E. and DICKEY, R. S. (1970). Histopathology of plants infected with vascular bacterial pathogens. *Annual Review Phytopathology*, 8, 259-80

PELCZAR, M. J., REID, R. D. and CHAN, E. C. S. (1977). *Microbiology*, McGraw-Hill, New York

SCHUSTER, M. L. and COYNE, D. P. (1974). Survival mechanisms of phytopathogenic bacteria. *Annual Review Phytopathology*, 12, 199-221

SKERMAN, V. B. D. (ed.) (1969). *Abstracts of Microbiological Methods*, Wiley-Interscience, New York

STANIER, R. Y., DONDOROFF, M. and ADELBERG, E. A. (1971). *General Microbiology*, Macmillan, London

STARR, M. P. (1959). Bacteria as plant pathogens. *Annual Review Microbiology*, 13, 211-38

STOLP, H. (1961). Neue Erkenntnisse über phytopathogene Bakterien und die von ihnen Verursachten zwischen Phytopathogenen *Pseudomonas* "Arten" und saprophytischen Fluoreszenten auf der Grundlage von Phagenreaktionen. *Phytopathologische Zeitschrift*, 42, 197-262

WILKINSON, J. F. (1982). *Introduction to Microbiology*, Blackwell, Oxford

FURTHER READING

Commonwealth Mycological Institute Descriptions of Pathogenic Fungi and Bacteria, published by Commonwealth Mycological Institute, Kew. Number: *Pseudomonas phaseolicola* (Halo Blight) 45

ANON (1966). International code of nomenclature of bacteria. *International Journal Systematic Bacteriology*, 16, 459–90

BRIAN, P. W. (1955). The role of toxins in the etiology of plant diseases caused by fungi and bacteria. 294–319, in *Mechanisms of Microbial Pathogenicity* (eds J. W. Howie and A. J. O'Hea), Society General Microbiology Symposium no. 5, Cambridge University Press, London

BUDDENHAGEN, I. W. (1965). The relation of plant-pathogenic bacteria to the soil. 269–84, in *Ecology of Soil Borne Plant Pathogens* (eds K. F. Baker and W. C. Snyder), John Murray, London

CROSSE, J. E. (1966). Epidemiological relations of the Pseudomonad pathogens of deciduous fruit trees. *Annual Review Phytopathology*, 4, 291–310

DAVIS, R. E. and WHITCOMB, R. F. (1971). Mycoplasmas, Rickettsiae and Chlamydiae: possible relation to yellows diseases and other disorders of plants and insects. *Annual Review Phytopathology*, 9, 119–54

DOWSON, W. J. (1957). *Plant Diseases due to Bacteria*, Cambridge University Press, London

GROGAN, R. G. and KIMBLE, K. A. (1967). The role of seed contamination in the transmission of *Pseudomonas phaseolicola* in *Phaseolus vulgaris*. *Phytopathology*, 57, 28–31

HAYWARD, A. C. (1974). Latent infections by bacteria. *Annual Review Phytopathology*, 12, 87–97

HOPKINS, D. L. (1977). Diseases caused by leaf-hopper borne Rickettsia-like bacteria. *Annual Review Phytopathology*, 15, 277–94

MOUNT, M. S. and LACY, G. H. (1982). *Phytopathogenic Prokaryotes*, Vol. I, Academic Press, New York

NELSON, P. E. and DICKEY, R. S. (1970). Histopathology of plants infected with vascular bacterial pathogens. *Annual Review Phytopathology*, 8, 259–80

NIENHAUS, F. and SIKORA, R. A. (1979). Mycoplasmas, Spiroplasmas and Rickettsia-like organisms as plant pathogens. *Annual Review Phytopathology*, 17, 37–58

PATEL, S. S. (1974). Toxins produced by phytopathogenic bacteria. *Annual Review Phytopathology*, 12, 259–79

PATON, A. M. (1960). The role of *Pseudomonas* in plant disease. *Journal Applied Bacteriology*, 23, 526–32

PÉROMBELON, M. C. M. and KELMAN, A. (1980). Ecology of the soft rot Erwinias. *Annual Review Phytopathology*, 18, 361–87

STAPP, C. (1961). *Bacterial Plant Pathogens*, Oxford University Press, London

STARR, M. P. (1959). Bacteria as plant pathogens. *Annual Review Microbiology*, 13, 211–38

STOLP, H., STARR, M. P. and BAIGENT, N. L. (1965). Problems of speciation of phytopathogenic Pseudomonads and Xanthomonads. *Annual Review Phytopathology*, 3, 231–64

WHITCOMB, R. F. (1980). The genus *Spiroplasma*. *Annual Review Microbiology*, 34, 677–709

VAN DER ZWET, T. and KEIL, H. L. (1979). Fire Blight—a Bacterial Disease of Rosaceous Plants, *United States Department of Agriculture Handbook* No. 510

5 CONTROL: HOST RESISTANCE

5.1 FORMS OF PATHOGEN CONTROL

Control of plant pathogens is sought by five main avenues: biological, chemical, cultural (including post-harvest control), host resistance and plant health regulation. Each of these forms of control aims at a different facet of the host–parasite–environment–time interaction. Resistance is directed at the host in an attempt to produce plants that are unaffected by or can withstand the pathogen *per se*; chemical control aims at the pathogen in an attempt to destroy it before or immediately after invasion and to prevent extensive colonisation and sporulation; cultural control aims at changing the environment that is encountered by the pathogen, either inside or outside the host, thereby providing conditions conducive to host growth but adverse to the pathogen; plant health regulation aims to prevent pathogens becoming endemic and to provide the producer with sources of healthy planting material; biological control aims to use negative interactions between micro-organisms such as competition, antibiosis or antagonism. Crop scientists have been largely preoccupied with the use of resistance, chemicals and plant health regulation. The grower has often used cultural controls with success but without any rationale for their operation. Since cultural controls are a multidimensional approach, the factors involved are less clearly drawn and less well comprehended, thereby lacking the conceptual cleanliness of other forms of crop protection and appealing less to those who seek to interpret disease control by the scientific method. This is an unfortunate situation because cultural controls, by utilising environmental changes, are a potent means of pathogen control. This can be all too evident when comparing disease incidence on well-husbanded holdings and on those that are poorly managed.

In the longer term all five forms of control are likely to be used in an integrated form in order to push the levels of return from crops still higher (section 9.1.6). In the short term, however, the main tool available when disease has developed or occurs with predictable regularity is the use of agrochemicals. Greatest improvements with these materials are most likely to come from more efficient and safer application methods that use less of the active chemicals. The costs of discovering, developing, manufacturing and registering agrochemicals are now so high that it is likely that fewer new materials will appear; consequently the need is for more effective use of those already available. Cultural controls will continue to be applied by growers and hopefully investigation of their *modus operandi* will lead to more general application. Resistance and plant health regulation are applied outside the holding; where possible the grower purchases seed of a resistant cultivar or plants from virus-tested stock. Biological control has yet to be exploited except for a few notable successes. The main purpose is, however, not to control plant pathogens, but to produce fruitful and profitable crops, which means that control measures must fit into the form of production schedule that the grower requires.

5.2 THE DESCRIPTION OF RESISTANCE

Resistance to pathogens is a genetically inherited character similar to other attributes such as height, yield and leaf size; but since two organisms are involved in a disease syndrome, host and pathogen, an ecological situation ensues in which the effectiveness of resistance depends on the characteristics of host and pathogen which are moderated by the environment in which they grow. The interplay of these factors may be described by the triangle

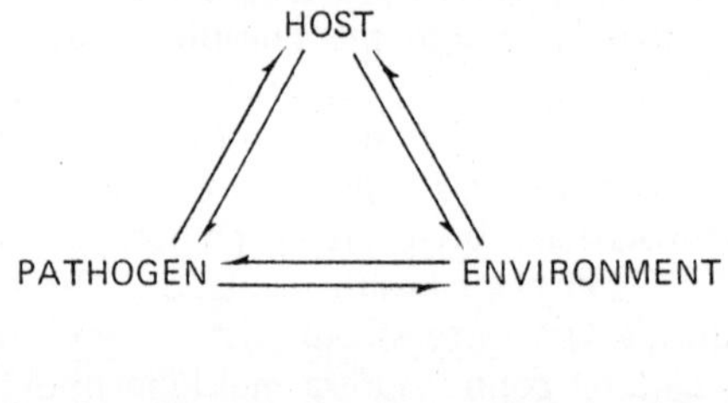

This is a simplification because for the triangle to be completely symmetrical the host and pathogen would each need to influence the environment to equivalent degrees and this is an unlikely event; also the dimension of time over which interaction takes place should be considered. Various terms are used to describe the interaction between plant and micro-organism.

Immunity

This is an absolute term indicating that there is no reaction at all between particular host and micro-organism species; thus the cabbage (*Brassica oleracea*) is not invaded by *Phytophthora infestans* (potato blight).

Resistance and susceptibility

These are relative terms indicating some degree of interaction between host and micro-organism species. Certain cultivars and wild varieties within a host species may be resistant to a particular pathogen, showing a low degree of interaction known as *non-compatibility*; the pathogen fails to establish an infection and subsequent disease symptoms. Susceptibility, on the other hand, indicates a high degree of interaction between host and micro-organism known as *compatibility*; the host becomes fully invaded and disease symptoms are manifested. A gradient between resistance and susceptibility can be identified within the host species and cultivars described according to the intensity of reaction as: *Very Resistant, Resistant, Moderately Resistant, Slightly Resistant, Susceptible and Very Susceptible*. These terms are not absolute and the expression of resistance is affected by the environment; for example, tomato cultivars possessing the gene Cf_1 for resistance to *Fulvia fulva* (leaf mould) are resistant if grown in long days with high light intensity but susceptible if grown in short days at low light intensity.

Tolerance

The interaction between host and pathogen is described as exhibiting tolerance when the host is invaded and disease symptoms are as severe as in comparable cultivars but the yield reduction (a usual parameter for measuring resistance) is lower.

Hypersensitive resistance

This term is used to indicate that host invasion takes place but the cells initially entered by the pathogen are so sensitive to infection that they collapse and die, forming a barrier that prevents further spread of the pathogen.

Disease escape

When a host is susceptible but because of spatial or time separation between host and pathogen no infection can take place, this is termed disease escape.

5.3 MODES OF INHERITANCE

The introduction of resistance to pathogens during this century has been a major feature in crop improvement by breeding. It was thought, initially, that introduction of resistance would provide a straightforward, built-in and economic crop-protection mechanism and since there are no environmental hazards involved with the use of resistance there are compelling reasons for its use. Variation within the pathogen population has meant, however, that particularly for aerially dispersed, foliar-invading pathogens there has been concomitant development of physiological pathogen races compatible with resistant cultivars which often have rapidly negated the value of such cultivars.

The simplest manner by which resistance is transferred is as single dominant or recessive genes inherited in a Mendelian manner.

Rr × Rr (where R = resistant; r = susceptible)

↓

F_1 RRrr (all progeny have a resistant phenotype)

↓ selfing

F_2 RR : Rr : Rr : rr (progeny in the ratio 3 resistant : 1 susceptible phenotype)

or

Rr × Rr (where R = susceptible; r = resistant)

↓

F_1 RRrr (all progeny have a susceptible phenotype)

↓ selfing

F_2 RR : Rr : Rr : rr (progeny in the ratio 3 susceptible : 1 resistant phenotype)

Resistance may be described further in genetic terms as follows.

Oligogenic resistance

This results from the presence of one or a few genes having readily demonstrable effects; for example, resistance to *Fulvia fulva* (leaf mould) in tomato cultivars is termed oligogenic or major gene. Usually such resistance is specific and most commonly inherited in a dominant manner although recessively inherited oligogenic resistance is known, as in peas to

Erysiphe pisi (powdery mildew). It is usually located for breeding programmes by screening collections of heterogeneous but related plants, at the seedling stage, to identify heritable resistance.

Polygenic resistance

This form of resistance operates through many genes that individually have small effects. It is often equated with field resistance whereby a particular cultivar or wild variety is less prone than others to severe disease symptoms but this feature is not apparent when screening tests are conducted on seedlings, for example, resistance of cv. Majestic potato to *Phytophthora infestans* (blight, section 2.6.2).

Cytoplasmic resistance

The cytoplasm and associated organelles are now accepted to have a heritable role and this includes a transfer of resistance characters from one generation to the next. This is illustrated by symptom expression in the progenies of interspecific crosses of *Capsicum* spp. when infected with potato virus X (PVX). The parent species *C. annuum* and *C. pendulum* display systemic mottle and necrotic local lesions respectively when infected by PVX. In hybrid progeny, symptoms depend on which species is used as the female parent, so that

C. annuum ♀ × *C. pendulum* ♂

↓

progeny display systemic mottle with PVX

C. pendulum ♀ × *C. annuum* ♂

↓

progeny display necrotic local lesions with PVX

5.4 THE EFFECT OF VARIATION WITHIN THE PATHOGEN POPULATION

A major problem confronting the plant breeder attempting to produce pathogen-resistant crops is that micro-organisms vary at the species and subspecies levels just as greatly, if not more so, than the higher plants into which he is attempting to introduce resistance genes. Mating systems in many fungi (the ability of fungi to overcome resistance in plants has been studied in more detail than for bacteria and viruses) may be regarded as means to promote outbreeding and therefore greater heterogeneity (section 2.2).

The first demonstration that a fungal species contained strains that differed in *pathogenicity* (the ability to cause disease) was by Eriksson (1894) who showed that *Puccinia graminis* (black stem rust) when isolated from wheat did not infect oats, rye or some grass species. Isolates from a range of hosts were characterised by the ability to infect certain host species but not others. This enabled Eriksson to define several subspecies of *P. graminis*. Subsequently fungal species and subspecies were shown to vary in their ability to infect single cultivars or wild varieties within a host species. This was first demonstrated by Barrus (1911) who distinguished two physiological races, α and β, of *Colletotrichum lindemuthianum* (anthracnose) that had differing pathogenicity towards cultivars of green bean (*Phaseolus vulgaris*). Subsequently it has been shown that there are many physiological races within Eriksson's subspecies of *P. graminis* capable of differentially infecting cultivars of wheat (Green, 1971).

Physiological races may be defined as morphologically and physiologically similar groupings within a fungal (or bacterial) species but characterised by being differentially capable of causing disease to cultivars within a host species. This leads to the concept of differential series of cultivars whereby physiological races may be identified. If for any particular set of physiological races one identifies two levels of pathological reaction (resistance or susceptibility) on the hosts, then n differential hosts whose reaction is governed by a single gene pair can differentiate 2^n physiological races. If six levels of pathological reaction can be recognised (section 5.2) then n hosts are potentially capable of differentiating 6^n races. For example, if 10 differential hosts were used then 6^{10} (60 466 176) physiological races could be identified. Clearly such a situation is untenable at the very least from the plant breeder's point of view and attempts are now being made to relate variation within pathogen isolates to the genetic characteristics of both pathogen and host.

Variation in the fungi is due to the wide variety of life cycles and patterns of mating that exist in them and provide mechanisms for the storage and release of variability. This release is through segregation and recombination following hybridisation between genetically unlike types. Species of pathogenic fungi are, in consequence, composed of many physiological variants having different pathogenic capabilities. Some variants may have no capacity to cause disease on particular cultivars, that is, they are avirulent whereas others may become highly virulent.

Recombination of genetic material during meiosis has long been recognised as a source of variation in the genetic constitution of living organisms. Some fungi, however, have only an asexual reproductive system but there is ample evidence that these are as flexible in their characteristics, including pathogenicity, as the sexually reproductive types. Operation of the parasexual cycle was described by Pontecorvo and Roper (1952) and Pontecorvo (1954) working with *Aspergillus nidulans* and provided an explanation for this flexibility. Plasmogamy, karyogamy and haploidisation (section 2.2.2) take place in the parasexual cycle but not at a specified time or place in the life cycle of an organism. The sequence of basic events in a complete parasexual cycle is as follows.

Formation of heterokaryotic mycelium

Many fungi, particularly Ascomycotina and Deuteromycotina, possess multinucleate thalli and hyphal fusions develop whereby nuclei move between different hyphae. Cells thereby possess genetically dissimilar nuclei and are termed heterokaryons. Fungi that are dependent solely on a sexual cycle as a source of variation pass through the processes of fertilisation, segregation and recombination before variant types develop. Heterokaryosis improves on the mitotic cycle as a source of variation and may supplement or replace the meiotic cycle.

Fusion between two nuclei

When a mycelium becomes heterokaryotic, nuclear fusion may take place between haploid nuclei of different genotypes as well as between nuclei of the same type. This gives rise to heterozygous diploid and homozygous diploid nuclei respectively. Consequently the mycelium may contain two types of haploid, two types of homozygous diploid and two types of heterozygous diploid nuclei. These will presumably multiply at equivalent rates but the proportion of diploids will be lower than haploids. Pontecorvo (1958) estimated one diploid to 1000 haploid nuclei.

Mitotic crossing-over

Readers will be familiar with crossing-over as a phenomenon found at the diplotene stage of meiosis. It is known that a similar process is found in diploid cells of micro-organisms and results in new genetic combinations and linkages. Mitotic crossing-over is probably the most important phase of the parasexual cycle for the development of variation.

Sorting-out of diploid strains

In those fungi producing uninucleate conidia, sorting-out of diploid nuclei occurs by their incorporation into conidia which germinate and produce diploid mycelia.

Haploidisation

Diploid colonies will often form sectors producing haploid conidia; hence some diploid nuclei undergo haploidisation in the mycelium and are sorted out. Random recombination takes place during the process by which a diploid nucleus forms two haploid nuclei. In this process all genes on a single chromosome act as though they were linked. Since the frequency of haploidisation is greater than that of fusions to form diploid nuclei, the ultimate result of the growth of a diploid heterozygote, which formed from fusion of unlike nuclei in a heterokaryon, is a variety of haploid strains recombining the chromosomes of the diploid in all possible ways.

The parasexual cycle may be summarised as follows.

1. Formation of heterokaryotic mycelium
2. Fusion between like nuclei and fusion between unlike nuclei
3. Multiplication of diploid and haploid nuclei
4. Crossing-over during mitotic multiplication of diploid nuclei
5. Sorting-out of diploid nuclei
6. Haploidisation of diploid nuclei
7. Sorting-out of new haploid strains

Bacteria exhibit pathogenic variability similar to that of the fungi and possible mechanisms for this are described in section 4.1.

5.5 HOST-PARASITE INTERACTION

Whether or not a cultivar exhibits resistance or susceptibility to a particular physiological race of a pathogen depends on its genotype for resistance and the genotype for virulence or avirulence on the part of the physiological race in question. Consequently the development of disease in the host involves interaction of genes conditioning resistance in the host and those conditioning virulence in the parasite. This is known as the host–parasite interaction and attempts to understand and explain this interaction usually involve the 'gene-for-gene-theory'. In general terms this theory may be stated as: 'the ability of a pathogen to grow and produce symptoms on a host carrying major genes for resistance is determined by alleles governing virulence at corresponding loci in the pathogen'. Thus for each gene in the host capable of mutating to give resistance, there exists in the pathogen a gene capable of mutating to overcome that resistance. This theory was developed as a means of understanding the resistance relationship between flax cultivars (*Linum usitatissimum*) and isolates of the rust fungus *Melampsora lini* (reviewed by Flor, 1971). Genes for reaction to rust infection in the host were identified by the virulence of specific races of the parasite and, conversely, genes for virulence in the parasite were identified by the reaction of specific cultivars of the host. Cultivars with the same reaction to all races were considered to possess identical genes for rust reaction. Races having the same virulence characteristics on a cultivar were considered to possess identical genes for virulence.

There are four possible interactions among a pair of alleles governing resistance in a host and a corresponding pair controlling virulence in the pathogen (Loegering and Powers, 1962) and this may be illustrated by what is termed the *quadratic check*

		host genotypes	
		R	r
pathogen	A	resistant	susceptible
genotypes	a	susceptible	susceptible

(*Note*: capital and lower-case letters differentiate alleles only and do not imply recessiveness or dominance.)

Use of the quadratic check assumes that the differences between resistant and susceptible reactions will be clearly defined and resistance is associated with the absence of disease symptoms or fungal reproductive structures. This situation can often be satisfied to an extent for foliar-invading airborne pathogens where lesion types may be arbitrarily classified as either resistant or susceptible but is less clearly defined with other types of host–parasite combination. The gene-for-gene theory has been a powerful tool in developing an understanding of such pathogens where numerous, apparently unrelated, physiological races have segregated as a result of the use by plant breeders of monogenic resistances in a series of host cultivars. Gene-for-gene relationships have been postulated for several host–parasite combinations in horticultural crops as shown in table 5.1.

For host–parasite combinations that are not of the airborne foliar-invading type, particularly where the host is of an out-breeding type giving rise to a high level of innate variation in the host population, the line between resistance and susceptibility is far more difficult to establish. In table 5.1 only two (marked *) soil-borne pathogen–host combinations are listed. A great deal of effort has gone into understanding host–parasite relations in pathogens that invade cereal crops, particularly those causing bunt, powdery mildew, rust or smut diseases and for these several more gene-for-gene relationships have been postulated.

5.6 VERTICAL AND HORIZONTAL RESISTANCE

So far, resistance has been discussed from a genetic viewpoint as either oligogenic or polygenic relating to the numbers of genes involved and the magnitude of their individual effects. When considering resistance from the standpoint of a host–parasite interaction it may be described as either vertical or horizontal (Robinson, 1969).

Table 5.1

Host-parasite combinations affecting horticultural crops for which a gene-for-gene relationship is postulated

Host	*Pathogen*	*Reference*
Malus	*Venturia inaequalis*	Boone and Keit (1957) Day (1960)
Solanum	*Phytophthora infestans*	Black *et al.* (1953) Toxopeus (1956)
Lycopersicon	*Fulvia fulva*	Day (1956)
Solanum	*Synchytrium endobioticum**	Howard (1968)
Lactuca	*Bremia lactucae*	Crute and Johnson (1976)
Brassica	*Plasmodiophora brassicae**	Toxopeus (1979)
Helianthus	*Puccinia helianthi*	Sackston (1962) Miah and Sackston (1970)
Leguminosae	*Rhizobium* spp.	Nutman (1969)
Lycopersicon	tomato mosaic virus	Pelham (1966)
Lycopersicon	spotted wilt virus	Day (1960)

*Soil-borne pathogens.

Vertical resistance relates to the oligogenic situation where only one or a few host genes operate to produce resistance to a particular pathogen and that pathogen is capable of segregating to form several physiological races. In other words, vertical resistance is very specific and effective only against a small portion of the total population of a particular pathogen species. Hence it is generally equated with the gene-for-gene theory and, since its inheritance is governed by a single or few genes, it is easy to manipulate in a breeding programme. For this reason it is often a preferred pathway to use when compared with other forms of resistance where they can be identified. Vertical resistance can provide complete control of a particular physiological race or group of races but it is liable to 'break down' when the pathogen population produces variants compatible with the major gene(s) used in the plant-breeding programme. This situation may be described in terms of the pathogen by saying that the frequency of virulences within the pathogen population has changed due to selection pressure placed on it by the presence of a vertically resistant host. Shifts in virulence frequency are thought to occur because, although a crop of a monogenically resistant host is apparently free from infection, there will be a small number of propagules within any population of the pathogen capable of causing infection albeit weakly. This is due to the normal processes of variation within any pathogen population. Because these propagules are able to multiply without competition, a shift occurs in the frequency of virulences within the pathogen population and apparently the host resistance 'breaks down'. In fact the resistance has not broken down; it remains the same but the pathogen has changed in its characteristics. This is illustrated in figure 5.1. The most important feature is

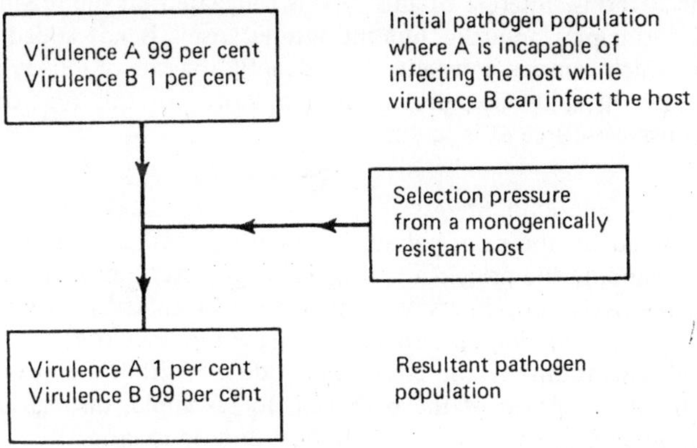

Figure 5.1 Changes in the virulence composition of a pathogen population as influenced by a resistant host

the rate at which such a change in virulence frequency takes place. If the change takes many years, then the use of monogenic resistance as a means of pathogen control is justified; if the change can take place within one or two seasons or crop cycles then use of such resistance is likely to fail in cost/benefit terms. The study of rates of pathogen spread, and concomitantly rates of virulence change taking place in a population, is embraced in the subject *epidemiology* which is broadly the study of factors affecting the outbreak of disease and spread of infectious pathogens. Assessments or quantification of such changes are made by studying the development of disease symptoms on host plants either in the field or under controlled conditions over a period of time, from which *rate (r)* of spread can be calculated. If on day zero nil disease is detected, on day one 50 per cent of the plants are affected and on day two 100 per cent are diseased, then the rate of spread is 50 per cent per day, that is, $r = 50$ per cent per day, or less clumsily, $r = 0.5$ per day or $r = 0.5$ day^{-1}. An infection rate of $r = 0.5$ is extremely rapid and indicates that the pathogen population is highly pathogenic, containing a high frequency of virulences compatible with host resistance if the host is vertically resistant. If the host is horizontally resistant (see below) the pathogen would be termed 'highly aggressive'. Generally *r* will be less than 0.5, thereby producing a curve rather than a straight line. The forms of curve involved have been equated with those obtained when money is invested under different interest systems.

Simple interest

In financial terms, interest of this type is calculated on the original sum invested (principal) annually but the interest itself is not added to the principal which remains constant. Consequently the rate of money growth is very slow. In epidemiological terms a pathogen of this type develops from a constant source of inoculum.

Discontinous compound interest

Interest earned by the principal sum of money is added to the principal thereby expanding the principal and its earning power but interest is only added at relatively long intervals of time such as annually so growth is still slow. In epidemiological terms the pathgoen inoculum increases with successive generations of the pathogen produced within a single season but the mode of spread of the pathogen, for example, through the soil precludes quick movement from host to host or crop to crop.

Continuous compound interest

Interest can be added to the principal and exert and earning power over short periods of time such as days, when large sums of money are lent on the world money markets, or monthly as with credit cards. Epidemiologically, the time between generations of pathogens of this type is very short and they spread rapidly, for example, by air currents. Consequently disease development is rapid and changes in virulence frequency are quickly reflected in the 'break down' of resistant cultivars (van de Plank, 1960, 1963).

The characteristics of vertically resistant hosts may be summarised as follows.

1. Perennial crops and those that are difficult to breed cannot be replaced with sufficient speed for vertical resistance to be of value if the pathogen is of the compound interest type.
2. Pathogen segregants with a new virulence combination spread only slowly if they are of the simple interest type but rapidly if of the compound interest type. Consequently resistance is more likely to be valuable against the former than against the latter.
3. Where a pathogen can produce a new virulence segregant quickly it is highly mutable and the use of vertical resistance should be avoided.
4. Crops that are grown as extensive monocultures are genetically

uniform or have large areas of the same cultivar exert a selection pressure on the pathogen which encourages segregation of new virulence combinations capable of attacking the crop.

5. If stabilising pressure through the use of a 'strong' gene for vertical resistance can be exploited, such resistance is likely to be of greater value. The strength of resistance genes is measured against the rate of decay of, or disappearance of, the equivalent virulence genes in the pathogen population. It is thought that with 'strong' resistance genes this rate of decay is faster than with 'weak' resistance genes. Stabilising pressure is the antithesis of selection pressure. Thus selection pressure from a particular form of host vertical resistance leads to an increase in the equivalent pathogen virulence. Withdrawal of a particular vertical resistance leads to a stabilising pressure on the pathogen population such that a specific virulence frequency may decrease, possibly to extinction.
6. If the pathogen is an obligate, at least two 'strong' genes are required to exploit stabilising pressure. This is because stabilisation can occur only in the parasitic growth phase where the parasite is growing on a host that lacks the complementary resistance gene. It is postulated that to exploit stabilising pressure in this way cultivars with different vertical resistance combinations must be grown in controlled patterns and for this at least two 'strong' genes for vertical resistance must be known. If the pathogen is a facultative parasite, stabilising pressure can occur also during its saprophytic growth and from this it is inferred that only one 'strong' gene is required.
7. It is thought that using mixtures of cultivars within a single season will prevent the build up and spread of new pathogen virulence combinations. This avenue of control is most likely to be effective against compound interest pathogens.
8. Using various combinations of host resistance in different seasons is more likely to be effective against simple interest pathogens.
9. Where the pathogen is transmitted with propagating material, for example, seed, vertical resistance is likely to have little value. It requires only a single infection focus in the crop from an infected seed or tuber for the pathogen to spread rapidly outwards especially if the pathogen is of the compound interest type.
10. A vertical resistance mechanism may confer either complete or incomplete protection against non-matching virulences. If the protection is incomplete the pathogen can be grown on the host and this increases the chances of segregation of virulences capable of overcoming the vertical resistance completely.
11. The value of vertical resistance is increased where for climatic or cultural reasons there is a complete break between the end of one crop and sowing of the next.

12. If legislation can be used to prevent the use of specific cultivars or crops it will enhance the life expectancy of vertical resistance.
13. Incorporation of horizontal resistance with vertical resistance is likely to provide more satisfactory control of a particular pathogen than vertical resistance alone. If a cultivar possesses only vertical resistance which is overcome by a pathogen the resultant host plants will be extremely susceptible and crop losses very high.
14. It is speculated that a host with complex vertical resistance, that is, containing several major genes, will suffer less disease when the resistance is overcome by a pathogen than a host with a single or few major genes for resistance (Robinson, 1971).

Horizontal resistance in a host implies possession of some measure of protection against the entire population of a particular pathogen. This may be tested by exposing a series of cultivars to a large number of isolates of the pathogen and if the ranking of cultivars for resistance/susceptibility remains constant, irrespective of the pathogen isolate used, then horizontal resistance is likely to be operative in those cultivars. Previously, horizontal resistance has been interpreted also as a form of tolerance and there is such considerable confusion in the literature concerning these terms that some authorities have suggested that the concept of tolerance should be abandoned. This confusion has arisen mainly in host–parasite combinations where parasite growth is reduced, for example by reduction of numbers of leaf lesions, or the sporulation within lesions, or the amount of mycelium colonising the vascular bundles is smaller in comparison with other combinations. These would now be classed as manifestations of horizontal resistance. It is unlikely that tolerance can provide a high level of disease protection and it can be used only in combination with other measures. Horizontal resistance is frequently equated with 'minor gene resistance' and its suggested mechanism is through the interactive effects of several to many genes that individually have small effect. They operate collectively to produce a cultivar with a measure of resistance. It must be accepted that because horizontal resistance is polygenically inherited its levels may differ in degrees from one genetic background to another and only rarely will it provide complete protection against a pathogen. Indeed the usefulness of attempting to breed for horizontal resistance in some horticultural crops may be questionable. This is especially so where the pathogen affects the organ (root, leaf, flower or fruit) that is eaten. The loss of product quality attendant on the use of horizontal resistance may be so great as to make it unacceptable to the consumer. It is possible, however, that with sufficient understanding of the mechanisms of horizontal resistance it may prove possible to combine it with high yields and quality.

Other factors also militate against horizontal resistance. In some crops, for example, use of high-density planting alone would be sufficient to negate the usefulness of horizontal resistance because under these conditions the pathogen is put at a highly favourable ecological advantage *vis-à-vis* the host, thereby preventing horizontal resistance from acting as an effective control. Changes in cultivation systems, such as the concentration of rotation, can give the pathogen an advantage over the horizontally resistant host. Nonetheless, vertical resistance has been proved to have so many attendant difficulties and to require such expensive continuous breeding programmes, that horizontal resistance should be tested to see whether it can be exploited more successfully, particularly in vegetatively propagated crops.

It is well recognised that some resistances have been maintained over considerable time periods and in the face of highly virulent new strains of a pathogen. Such resistance has now been termed *durable*. This term can be employed without knowledge of its mechanism, whether the genetic control is race specific or not, thereby implying no necessity for knowledge of the racial composition of the pathogen population.

5.7 SOURCES OF RESISTANCE

Resistance in crop plants is derived by plant breeders from a variety of sources.

1. *From within the same species* either by using cultivars with known resistance or from wild varieties and subspecies which are tested and found to possess resistance.
2. *From other species within the same genus* showing resistance to a pathogen.
3. *From other genera*: occasionally resistance from this source may be used where intergeneric hybrids can be developed that are either fertile and set seed or where once a hybrid is formed further multiplication may be made by vegetative propagation. In this type of breeding, future crosses may be achieved by 'genetic engineering' where it is thought that parts of chromosomes or even whole genomes may be transferred from one cell to a completely foreign cell and still exert a genetic effect. Such work is only in its infancy but the prospects for breeding pathogen-resistant plants are being widely recognised.
4. *Mutations of chromosomes or whole genomes* either naturally or stimulated by radiation or mutagenic chemicals. This is another avenue whereby pathogen-resistant crop plants are being developed. The use of these techniques is outlined in the following review of the use of pathogen resistance in horticultural crops.

5.8 USE OF RESISTANCE IN HORTICULTURAL CROPS

Resistance is used as a means to control losses caused by plant pathogens in all groups of horticultural crops. It is perhaps most widely applied and the genetic mechanisms responsible for its operation best understood with vegetable crops, followed by top and soft fruits, while in ornamental crops little fundamental knowledge is available. This situation is being corrected particularly by research on forest trees. Such work has immediate ramifications for arboricultural species. In addition, greater general availability of leisure time is leading to far higher financial values for ornamental material and concomitant demands for research to improve the quality of such stock. One of the main reasons for loss of quality is pathogen attack; hence studies of resistance are of increasing importance for these crops.

Vegetable crops present several good examples of the use of pathogen resistance that has remained effective over long periods of time. This is especially so with soil-borne pathogens where the rate of spread is relatively slow compared with airborne organisms and consequently the rate of change in virulence frequency is low. These pathogens are in consequence simple interest types. Resistance in brassicas to the soil-borne organism *Fusarium oxysporum* f. sp. *conglutinans* (cabbage yellows) has remained stable for decades. Initially, resistant lines were obtained by field selection; this led to cultivars possessing an incomplete resistance which at high temperatures could be overcome by the pathogen. Subsequent use of inbreeding techniques produced more resistant types. Eventually two forms of resistance were recognised: Type A is monogenically, qualitatively inherited and stable at high temperatures, Type B is polygenically, qualitatively inherited and gives incomplete resistance (Walker, 1959). A similarly successful case of stable resistance is that of onion to *Colletotrichum dermatium* f. sp. *circinans* (onion smudge); here resistance is related to the colour of the onion bulb scales the inheritance of which is well described and use of dark-skinned cultivars has led to almost total freedom from disease (Angel *et al.*, 1930; Link and Walker, 1933).

Aerial organ-invading pathogens of vegetable crops present several cases where many new cultivars have been bred and concomitantly new pathogen physiological races have been found, such as the bacterial pathogen of green beans, *Pseudomonas phaseolicola*, which causes halo blight (section 4.5). A more complex situation is found with the fungal pathogen of green beans which causes anthracnose (*Colletotrichum lindemuthianum*). Here a complex list of races has been subdivided into race groupings alpha to iota (Fouilloux, 1976). In the host there is a similarly complex series of resistance genes with 8–10 genes grouped in up to three allelomorphic series containing both duplicate and complementary genes for resistance; some genes are recessive in their mode of operation. At least three independent genes, for instance, will provide resistance to race delta. As with *P. phaseolicola* a single source of resistance to many races of *C. lindemuthianum* has been identified and is known to be due to a single gene ARE. Perhaps unwisely this single resistance source is now being used widely by virtue of its convenience in breeding programmes.

There are some classic compound interest types of host-parasite combinations to be found in vegetable crops. In the lettuce - *Bremia lactucae* (downy mildew—section 2.6.3) combination, some 200 resistant cultivars are known and about 30 physiological races have been identified. Most cultivars are derived from a narrow genetic base of less than ten resistance genes; for glasshouse types in particular resistance breeding has rested on the use of three or four genes. This situation largely resulted from intense competition on the part of plant-breeding companies to produce new cultivars regularly with little thought given to the consequent generation of new physiological races of *B. lactucae*. It was barely surprising, therefore, that new resistant cultivars were attacked by complementary virulences within 6 to 24 months of release into widespread commercial use. It may now be necessary to use interspecific hybrids in order to find resistance to *B. lactucae* but even this avenue is questionable since initial use of crosses with *L. serriola* was quickly matched by compatible virulences in the pathogen population. The use of interspecific crosses with *L. saligna* and *L. virosa* is being investigated together with horizontal resistance which may be available in some *L. sativa* cultivars where the rate of invasion by *B. lactucae* is slower (Crute and Dixon, 1981).

A comparable situation exists with glasshouse vegetables such as the cucumber and powdery mildew disease caused by *Sphaerotheca fuliginea*. Considerable breeding effort was put in to the use of monogenic dominant resistance only to be accompanied by an evolution of compatible pathogen races. Resistance was drawn from cantaloupe melon *Cucumis melo* var. *cantalupensis*; the host genes used were Pm^1—a single dominant, Pm^1 Pm^2—partly dominant but accompanied by several modifying genes, Pm^3—single dominant, Pm^4—partially dominant and Pm^5—completely dominant (Harwood and Markarian, 1968). In the cucumber itself (*Cucumis sativus*) a major recessive gene s determines resistance to powdery mildew in the hypocotyl and is of an intermediate type but essential for complete leaf resistance. This is further controlled by a single dominant resistance gene that is only expressed in the presence of gene s. Additionally gene I is known which is an inhibitory gene preventing full expression of resistance but does not affect gene s.

One of the earliest gene-for-gene relationships was postulated for the host-parasite combination tomato-*Fulvia fulva* (leaf mould). The large numbers of physiological races may be related to the host resistance genes Cf_1 to Cf_5. These genes are obtained from *Lycopersicon esculentum* and in addition wild relatives such as *L. chiloense*, *L. hirsutum* and *L. pimpinellifolium*. The failure of interspecific crossing to control this compound interest pathogen may call into question this policy with other comparable pathogens such as *Bremia lactucae*. In tomato the various resistance genes are manifested phenotypically in different ways: Cf_1 produces a large chlorotic spot devoid of a sharp boundary which later becomes necrotic, Cf_3 has a similar effect but the chlorotic and later necrotic areas are somewhat less extensive, whereas Cf_2 and Cf_4 cause pinpoint necrosis.

The major world-wide pathogen of apple is *Venturia inaequalis* (scab—section 2.6.5). Two forms of host resistance are available: monogenic and polygenic. The best solution would be a combination of both. There is a range of resistance from highly susceptible to strongly resistant in apple cultivars and a series of pathogen biotypes. Strongly resistant cultivars are resistant to several biotypes but this is conditioned by environment and therefore unreliable. The cv. Antonowska has more reliable polygenic resistance which is claimed to give field resistance to a range of isolates. This type of resistance has been used mainly in Europe. Breeding using monogenic resistance has been done in the USA. This began in Illinois with a cross *M. floribunda* x cv. Rome Beauty; *M. floribunda* carries the resistance gene Vf. Progeny from this cross have formed the basis of breeding in the USA, Canada and, to some extent, Europe. Other species used to provide monogenic resistance include *M. micromalus*, *M. atrosanguinea* and *M. prunifolia*. Resistance is dominant and cultivars with this form of resistance such as Prima and Priscilla are now in trials. There is a large pool of scab-resistance single genes in *Malus*; six resistance loci are recognised and five physiological races of *V. inaequalis*.

Control of fire-blight caused by the bacterium *Erwinia amylovora* would be of major significance particularly for the pear industry. Cultivars derived from chinese sand pears (*Pyrus pyrifolia*) by interspecific hybridisation with *P. communis* (European pear), for example Le Conte, Kieffer and Garber, are more resistant than those from *P. communis* alone but suffer from fruit quality defects. In the USA screening tests of large numbers of species, species hybrids and cultivars have identified sources of resistance. In the UK breeding is directed to produce cultivars that form no second flush of bloom since this is a major point of entry for *E. amylovora*. Use is also made of resistant back-cross derivatives of *P. ussuriensis* x *P. communis* from the North American programme by crossing to high quality *P. communis* cvs. Host resistance expression is affected by the age, vigour, succulence and type of tissue infected, temperature and humidity relationships during the pre-infection and post-infection period, inoculum purity and concentration, virulence of the isolate and the method of inoculation. It is thought that similar genes for resistance occur in most sources of 'wild' resistance and that both polygenic and monogenic resistances operate. An additional complication in the search for resistance is that shoot resistance/trunk susceptibility and vice versa have been noted in, for example, cv. Magness.

Many physiological races of *Phytophthora fragariae* (strawberry red core) have been identified: USA (10), UK (11), Canada (6), Japan (6). Field resistance and race-specific resistance appear not to be correlated but field resistance is correlated with a low percentage of roots exhibiting 'red stele' symptoms. 'Aberdeen' parentage contributes race-specific resistance and 'Frith' parentage contributes race-specific and field resistance. Resistance is partially dominant and quantitatively inherited. Sources of resistance to USA races are: Aberdeen—an American domestic cultivar of unknown parentage, Md 683—a cultivated type containing Scottish parentage (Scotland BK 46) and probably derived from Frith, selected clones of Little Scarlet and selected clones of North American *Fragaria chiloensis*. Resistant cultivars include Early Midway, Gorella, Guardian, Hood, Midway, Redchief, Redgauntlet, Redglow, Shuksan, Sparkle, Sunrise, Surecrop and Talisman.

In raspberry (*Rubus idaeus*) it is possible to correlate resistance to some pathogens with morphological characters of the canes. *Didymella applanata* (spur blight) is less frequent on spine-free, wax-free and non-pigmented canes, while *Botrytis cinerea* (grey mould—section 2.6.4) increases slightly on wax-free canes. The cvs Lloyd George and Burnetholm are heterozygous for three genes governing resistance to *Sphaerotheca macularis* (powdery mildew). Genes Sp_1 and Sp_2 are dominant complementaries and sp_3 a recessive. Resistance is epistatic to susceptibility whether of Sp_1, Sp_2 or sp_3 origin. Gene sp_3 is linked to gene T for fruit colour with a cross-over value of 25 per cent. The gene order in this linkage group is designated as B-Sx_3-T-sp_3 and there is evidence of linkage to H gene (hairy canes). Inheritance of resistance to viruses such as raspberry ringspot (section 3.5.4), Arabis mosaic and tomato black ringspot is dominant to susceptibility for each virus but more than one gene is involved (Janick and Moore, 1975). The presence of resistance to some pathogens of raspberry in *Rubus* cultivars and species is summarised in table 5.2.

Studies of forest and ornamental tree pathogens have demonstrated the presence of pathogen physiological races in for example the host–parasite combinations: *Populus* and *Larix–Melampsora*, *Pinus–Melampsora* and *Pinus–Cronartium* (section 2.6.8). A gene-for-gene relationship is postulated in *Pinus monticola–Cronartium ribicola*, where three independent host loci interact with two independent pathogen loci; one of the pathogen loci has alternate alleles interacting with two independent host loci to produce four physiological races on *Pinus* foliage. Periodic 'breakdowns' in resistance, as has happened in cereals and potatoes would be even more disastrous in forest crops. The protracted reproductive cycles

Table 5.2

The presence of resistance to some pathogens in Rubus *cultivars and species (after Janick and Moore, 1975)*

Character	*Cultivar*	Rubus *species*
Spur blight resistance (*Didymella applanata*)	Ottawa Newburgh Marcy Pyne's Royal Preussen	*R. idaeus* *R. occidentalis*
Mildew resistance (*Sphaerotheca macularis*)	Malling Promise Lloyd George	*R. idaeus*
Mildew susceptibility (*Sphaerotheca macularis*)	Latham Carnival Ottawa	*R. idaeus* var. *strigosus*
Anthracnose resistance (*Gloeosporium allantosporum*)	Ottawa	*R. coreanus* *R. biflorus* *R. kuntzeanus* *R. nivens* *R. innominatus* *R. idaeus*
Verticillium resistance (*Verticillium albo-atrum*)	Cuthbert	*R. idaeus* var. *strigosus* *R. biflorus*
Cane spot resistance (*Elsinoë veneta*)		*R. coreanus* *R. glaucus*
Soil-borne virus resistance	Lloyd George Malling Landmark	*R. idaeus*
Green mottle resistance	Latham Chief	*R. idaeus* var. *strigosus*

of trees (5–25 years) and long forestry crop rotations (20–100 years) impose severe limitations on the tree breeder's use of, or reliance on, disease-resistance breeding schemes as used in other spheres of agriculture and horticulture. Attention is being directed towards co-evolved and balanced host to pathogen systems in centres of origin or diversity. The equilibrium in such centres is explained as 'reciprocal adaptation of antagonistic organisms (that) may, under selective pressures, eventually involve host and parasite in a definite progressive correlation, until a certain pathological balance is achieved in successive generations' (Bingham *et al.*, 1971).

A good example of the pathogen problems that confront breeders of forest trees and ornamental stock is that of Dutch elm disease caused by *Ceratocystis ulmi*. It was first recognised in North-eastern Europe and the UK during and after 1917. It spread throughout European elms and by 1929 had reached the Middle East. In this period *C. ulmi* also entered the USA and by 1933 was causing extensive damage throughout New Jersey. In the Netherlands 95 per cent of elms were derived from a highly susceptible clone of *Ulmus hollandica* var. *Belgica* and losses due to *C. ulmi* exceeded 70 per cent. In the north-eastern USA losses of susceptible *U. americana* were up to 15 per cent per annum in some cities but in the UK losses were relatively slight. The sudden appearance in 1917 of *C. ulmi* as a virulent pathogen may have been due to mutation of a saprophytic strain of *C. ulmi* and hybridisation between *Ceratocystis* spp. Similar questions of origin surround the emergence of a virulent strain of *C. ulmi* which since the early 1970s has decimated UK elms. Virulence in *C. ulmi* towards some elm clones is thought to be polygenically inherited and matched by a polygenic inheritance of resistance in certain elm clones. Polygenic inheritance of resistance has been equated in theory with stability of resistance but recent experience in the UK with elm suggests that polygenic resistance may 'break down' in just as spectacular a fashion as monogenic resistance.

Physiological races are also known to occur with host–parasite combinations of other ornamentals; for instance two races (1 and 2) of *Fusarium oxysporum* f. sp. *dianthi* affect *Dianthus* spp. Within these physiological races considerable interisolate variation for 'aggressiveness' (virulence) has been identified. There is also evidence of highly specific host-to-isolate interaction. Most carnation cultivars of the spray and miniature types are susceptible to *F. oxysporum* f. sp. *dianthi* race 1 whereas cv. Sim and its sports may be resistant. The reverse situation is found for race 2 except for cv. Orchid Beauty, a standard glasshouse carnation resistant to race 2. Crossing-programmes are utilising resistant cultivars such as Carrier 929, Maj Britt and Orchid Beauty together with partially resistant cvs Alice, Heidi, Silvery Pink and Tony (Matthews, 1979; Matthews *et al.*, 1979).

In the rose it is possible to rank cvs by field observation for resistance to *Diplocarpon rosae* (black spot). Hybrid Tea (large-flowered bush) cvs Pink Favourite and Silver Jubilee are most resistant while Troika and Wendy Cussons are most susceptible; floribunda (cluster-flowered bush) cvs Queen Elizabeth and Southampton are most resistant while Pink Parfait, Evelyn Fison, Matangi and Stephen Langdon are most susceptible (Hollis, 1981). Also rose species such as *Rosa bella* are recognised as highly resistant to *D. rosae* but *R. arvensis* is highly susceptible (Roberts *et al.*, 1981). These levels of resistance–susceptibility vary widely between seasons and localities.

In vegetable crops there is a well-documented case of changes in pathogenicity to host species. *Erysiphe cichoracearum* causes powdery mildew on lettuce, artichoke, chicory and endive. In California, USA it now infects both wild lettuce (*Lactuca serriola*) and the cultivated *L. sativa*. Initially this pathogen appeared only on *L. serriola* but later the asexual stage developed on *L. sativa*. More recently, mixed strains were found on *L. serriola*, suggesting that a mutant capable of infecting *L. sativa* had developed, because the sexual form was confined to *L. serriola*. Eventually perithecia extensively affected *L. sativa* cv. Great Lakes. This pathogen appears to have evolved from infecting only *L. serriola* through an asexual form that could attack *L. sativa*, and finally to have produced a biotype capable of carrying out its complete life cycle on *L. sativa*. Crisp types appear to be highly susceptible but some resistance is known within other types such as Arctic King, Big Boston, Salad Bowl and Bath Cross (Dixon, 1978).

Finally it must be recognised that some pathogens are so highly variable that strong resistance is unlikely to be found. For instance, *Botrytis cinerea* (grey mould—section 2.6.4) is a classically highly variable heterokaryotic pathogen attacking a wide host range but it may be possible to accumulate polygenes for resistance by field selection in certain crops such as green beans. Similarly *Ustilago maydis* (maize smut) is also complexly heterothallic, producing many biotypes; attempts to use monogenic resistance have failed but again accumulation of polygenes could yield low to moderate levels of resistance.

REFERENCES

ANGEL, H. R., WALKER, J. C. and LINK, K. P. (1930). The relation of protocatechuic acid to disease resistance in the onion. *Phytopathology*, **20**, 431

BARRUS, M. F. (1911). Variation in varieties of beans in their susceptibility to anthracnose. *Phytopathology*, **1**, 190-5

BINGHAM, R. T., HOFF, R. J. and MCDONALD, G. I. (1971). Disease resistance in forest trees. *Annual Review Phytopathology*, **9**, 433-52

BLACK, W., MASTENBROEK, C., MILLS, W. R. and PETERSON, L. C. (1953). A proposal for an international nomenclature of races of *Phytophthora infestans* and of genes controlling immunity in *Solanum demissum* derivatives. *Euphytica*, **2**, 173-9

BOONE, D. M. and KEITT, G. W. (1957). *Venturia inaequalis* (Cke.) Wint. XII. Genes controlling pathogenicity of wild-type lines. *Phytopathology*, **47**, 226-33

CRUTE, I. R. and DIXON, G. R. (1981). Downy mildew diseases caused by the genus *Bremia* Regel. 423-60, in *The Downy Mildews* (ed. D. M. Spencer), Academic Press, London

CRUTE, I. R. and JOHNSON, A. G. (1976). The genetic relationship between races of *Bremia lactucae* and cultivars of *Lactuca sativa*. *Annals Applied Biology*, **83**, 125-37

DAY, P. R. (1956). Race names of *Cladosporium fulvum*. *Tomato Genetics Cooperative Report*, **6**, 13-14.

DAY, P. R. (1960). Variation in phytopathogenic fungi. *Annual Review Microbiology*, **14**, 1-16

DIXON, G. R. (1978). Powdery mildews of vegetable and allied crops. 495-524, in *The Powdery Mildews* (ed. D. M. Spencer), Academic Press, London

ERIKSSON, J. (1894). Uber die Specialislrung des Parasitismus bei den Getreiderostpilzen. *Berichte Deutschen Botanischen Gesellschaft*, **12**, 292-331

FLOR, H. H. (1971). Current status of the gene-for-gene concept. *Annual Review Phytopathology*, **9**, 275-96

FOUILLOUX, G. (1976). L'anthracnose du Haricot (*Colletotrichum lindemuthianum* Sacc. et Magn.) nouvelles sources de resistance et nouvelles races physiologiques. *Annales Amelioration Plantes*, **26**, 443-53

GREEN, G. J. (1971). Physiologic races of wheat stem rust in Canada from 1919 to 1969. *Canadian Journal Botany*, **49**, 1575-88

HARWOOD, R. R. and MARKARIAN, D. (1968). The inheritance of resistance to powdery mildew in the cantaloupe variety Seminole. *Journal Heredity*, **59**, 126-30

HOLLIS, L. (1981). Hybrid teas and floribundas that resist blackspot. *Rose Annual*, Royal National Rose Society, London, pp. 33-9

HOWARD, H. W. (1968). The relation between resistance genes in potatoes and pathotypes of potato-root eelworm (*Heterodera rostochiensis*), wart disease (*Synchytrium endobioticum*) and potato virus X. *Abstracts First International Congress Plant Pathology*, London, p. 92

JANICK, J. and MOORE, J. N. (1975). *Advances in Fruit Breeding*, Purdue University Press, Indiana

LINK, K. P. and WALKER, J. C. (1933). The isolation of catechol from pigmented onion scales and its significance in relation to disease resistance in onions. *Journal Biological Chemistry*, **100**, 379-83

LOEGERING, W. Q. and POWERS, H. R. (1962). Inheritance of pathogenicity in a cross of physiological races 111 and 36 of *Puccinia graminis* f. sp. *tritici*. *Phytopathology*, **52**, 547-54

MATTHEWS, P. (1979). Variation in *Fusarium* wilt isolates of the carnation. *Annual Report John Innes Institute, Norwich, for 1978*, 42

MATTHEWS, P., ARTHUR, A. E., HOSIER, J. and BULLEN, D. (1979). Breeding carnations for resistance to *Fusarium* wilt. *Annual Report John Innes Institute, Norwich, for 1978*, 40-1

MIAH, M. A. and SACKSTON, W. E. (1970). Genetics of pathogenicity in sunflower rust. *Phytoprotection*, **51**, 17-35

NUTMAN, P. S. (1969). Genetics of symbiosis and nitrogen fixation in legumes. *Proceedings Royal Society London*, **B172**, 417-37

PELHAM, J. (1966). Resistance of tomato to Tobacco Mosaic Virus. *Euphytica*, **15**, 258-67

PONTECORVO, G. (1954). Mitotic recombination in the genetic systems of filamentous fungi. *Caryologia*, supplement, **6**, 192-200

PONTECORVO, G. (1958). *Trends in Genetic Analysis*, Columbia University Press, New York

PONTECORVO, G. and ROPER, J. L. (1952). Genetic analysis without sexual reproduction by means of polyploidy in *Aspergillus nidulans*. *Journal General Microbiology*, **6**, vii (abstract)

ROBERTS, A. V., CASTELDINE, P. and GRANT, B. W. W. (1981). Resistance and susceptibility to blackspot. *Rose Annual*, Royal National Rose Society, London, pp. 117-21

ROBINSON, R. A. (1969). Disease resistance terminology. *Review Applied Mycology*, **48**, 593-606

ROBINSON, R. A. (1971). Vertical resistance. *Review Plant Pathology*, **50**, 233-9

SACKSTON, W. E. (1962). Studies on sunflower rust. III. Occurrence, distribution and significance of *Puccinia helianthi* Schw. *Canadian Journal Botany*, **40**, 1449-58

TOXOPEUS, H. J. (1956). Reflections on the origin of new physiological races of *Phytophthora infestans* and the breeding of resistance in potatoes. *Euphytica*, 5, 221-37

TOXOPEUS, H. (1979). Preliminary proposal for a notation of clubroot resistance and pathogen virulence to explain E C D test results in terms of a 'gene-for-gene' hypothesis. 67-72, in *Proceedings of Woronin + 100 Conference 1977*, Wisconsin University, Madison

VAN DER PLANK, J. E. (1960). Analysis of Epidemics. 229-89, in *Plant Pathology—An Advanced Treatise. III. The Diseased Population, Epidemics and Control*, (eds J. G. Horsfall and A. E. Dimond), Academic Press, New York

VAN DER PLANK, J. E. (1963). *Plant Diseases: Epidemics and Control*, Academic Press, New York

WALKER, J. C. (1959). Progress and problems in controlling plant disease by host resistance. 32-41, in *Plant Pathology Problems and Progress, 1908-1958* (eds C. S. Holton, G. W. Fischer, R. W. Fulton, H. Hart and S. E. A. McCallan), Wisconsin University Press, Madison

FURTHER READING

ALLARD, R. W. (1960). *Principles of Plant Breeding*, John Wiley, London

DAVIS, R. H. (1966). Mechanisms of inheritance. 2. Heterokaryosis. 567-88, in *The Fungi* (eds G. C. Ainsworth and A. S. Sussman), Vol. II, Academic Press, New York

DAY, P. R. (1974). *Genetics of Host-Parasite Interaction*, Freeman, San Francisco

DIXON, G. R. (1981). *Vegetable Crop Diseases*, Macmillan, London

FINCHAM, J. R. S. and DAY, P. R. (1965). *Fungal Genetics*, Blackwell, London

HORSFALL, J. G. and DIMOND, A. E. (eds), *Plant Pathology—An Advanced Treatise*, vol. 1. The Diseased Plant (1959). vol. 2. The Pathogen (1960). vol. 3. The Diseased Population, Epidemics and Control (1960), Academic Press, New York

JINKS, J. L. (1952a). Heterokaryosis: a system of adaptation in wild fungi. *Proceedings Royal Society, London*, B140, 83-99

JINKS, J. L. (1952b). Heterokaryosis in wild *Penicillium*. *Heredity*, 6, 77-87

NELSON, R. R. (ed.) (1973). *Breeding Plants for Disease Resistance—Concepts and Applications*. Pennsylvania State University Press

NELSON, R. R. (1978). Genetics of horizontal resistance in plant diseases. *Annual Review Phytopathology*, 16, 359-78

NORTH, C. (1979). *Plant Breeding and Genetics in Horticulture*, Macmillan, London

PONTECORVO, G. (1956). The parasexual cycle in fungi. *Annual Review Microbiology*, 10, 393-400

ROBINSON, R. A. (1973). Horizontal resistance. *Review Plant Pathology*, 52, 483-501

ROBINSON, R. A. (1976). *Plant Pathosystems*, Springer-Verlag, Berlin

ROBINSON, R. A. (1980). New concepts in breeding for disease resistance. *Annual Review Phytopathology*, 18, 189-210

ROPER, J. A. (1966). Mechanisms of inheritance. 3. The parasexual cycle. 473-512, in *The Fungi—An Advanced Treatise*. vol. II (eds G. C. Ainsworth and A. S. Sussman), Academic Press, New York

SCOTT, P. R. and BAINBRIDGE, A. (eds) (1978), *Plant Disease Epidemiology*, Blackwell, London

STAKMAN, E. C. (1914). A study in cereal rusts, physiological races. *Minnesota Agricultural Experimental Station*, St. Paul, Bulletin No. 138

STAKMAN, E. C., PARKER, J. J. and PIEMEISEL, F. J. (1918). Can biologic forms of stem rust on wheat change rapidly enough to interfere with breeding for rust resistance? *Journal Agricultural Research*, 14, 111-23

THOMPSON, J. N. and THODAY, J. M. (eds) (1979). *Quantitative Genetic Variation*, Academic Press, London

VAN DER PLANK, J. E. (1968). *Disease Resistance in Plants*, Academic Press, New York

VAN DER PLANK, J. E. (1975). *Principles of Plant Infection*, Academic Press, New York

VAN DER PLANK, J. E. (1978). *Genetic and Molecular Basis of Plant Pathogenesis*, Springer-Verlag, Berlin

WATTS, L. (1980). *Flower and Vegetable Plant Breeding*, Grower Books, London

WILLIAMS, W. (1964). *Genetic Principles and Plant Breeding*, Blackwell, London

ZADOKS, J. C. and KOSTER, L. M. (1976). A historical survey of botanical epidemiology. A sketch of the development of ideas in ecological phytopathology. *Mededelingen Landbouwhogeschool Wageningen*, No. 76-12

ZADOKS, J. C. and SCHEIN, R. D. (1979). *Epidemiology and Plant Disease Management*, Oxford University Press, London

6 CONTROL: CHEMICALS

6.1 GENERAL PRINCIPLES

There are four main avenues for the chemical control of plant pathogens.

1. Protection of healthy hosts from infection (*prophylaxis*).
2. Cure or therapy of diseased plants (*chemotherapy*).
3. Destruction of pathogen propagules and resting stages present on the surfaces of dormant host tissue such as seeds or on the surfaces and in crevices of structures such as glasshouses (*disinfestation*).
4. Elimination of pathogens present in growing media such as soils and composts (*fumigation or sterilisation*).

These terms equate with those used to describe the chemicals employed

prophylaxis	–	protectant	chemical
chemotherapy	–	eradicant or curative	
disinfestation	–	disinfectant	
sterilisation	–	sterilant/fumigant	

The protective value of a compound is measured by its ability to prevent infection. To be most efficient it must be present in a high enough concentration of active material when pathogen inoculum arrives. Consequently it is necessary to determine the stage at which a host is most susceptible to infection, the environment likely to be most conducive to infection and when there is the greatest likelihood of invasion. Until about 1965 protective chemicals were applied only if there was a strong regular possibility of invasion. Since that date systemic chemicals (those capable of being absorbed into a host and of moving away from the site of application) have become available and protective spray programmes have been used more frequently and for a wider range of crops. This is because systemic chemicals are more reliable, have greater persistence, require less frequent application and protect new growth as it forms. In general the properties now required of a protectant are stability and retention either outside or inside a host. Eradicants are used to control pathogens after infection has started; thus they are applied once visible lesions or pathogen-growth is evident. Some systemic materials have both protectant and eradicant properties. Disinfectants are mainly used for seed treatments and sterilants or fumigants, which are mostly low boiling-point liquids or gases, are introduced into soil or compost to kill existing pathogens. None of these terms is absolute because an individual chemical may have more than one function. Chemicals used to control fungi are of three types.

1. *Fungicidal*—killing fungal tissue.
2. *Fungistatic*—inhibiting fungal growth.
3. *Genestatic*—inhibiting fungal sporulation.

There are many more compounds available commercially to control fungal pathogens than to control bacterial pathogens. Indeed the chemical control of plant-pathogenic bacteria has progressed little during the past 50 years. Bacteria may be seed-borne, perennate on vegetative propagation material and be spread by insect vectors or by mechanical transmission (section 4.3). The most satisfactory methods of control are, therefore, either by seed-certification schemes to maintain freedom from pathogens in original planting material (chapters 8 and 9) or by breeding resistant cultivars (chapter 5). There are no materials available commercially that directly control viruses, viroids or mycoplasma-like organisms. Control rests solely on reduction or elimination of vectors and in the prevention of reinfestation of pathogen-free planting material (chapters 8 and 9).

For a crop-protection chemical to be successful it must fulfil the following criteria.

1. It must provide adequate and consistent pathogen control.
2. Interaction with weather, soil and crop cultivars should be as limited as possible.
3. There must be no phytotoxic effects to the host crop at those dose rates necessary to provide protection.
4. Harmful effects on the environment, toxicity to humans, domesticated and wild animals and plants must be either non-existent or of an acceptably low level.
5. The active ingredient must be capable of formulation into a suitable physical form that is safe, easy and cheap to store, transport and apply.
6. Application must yield an adequate economic return for a given effort so that financial improvements from enhanced yield and/or quality exceed the total cost of control.

Factors that govern a grower's choice of chemical when the above criteria are achieved will be based on selection of the cheapest effective product at the minimum necessary dosage with the least number of treatments and most convenient method of application (Evans, 1977).

6.2 CHEMICAL NOMENCLATURE

Various terms are used to describe crop-protection chemicals.

Molecular formula
This is a shorthand description of those chemical elements and numbers of atoms of each element that compose a chemical compound.

Structural formula
This a pictorial diagram of the manner in which atoms within a compound are linked to each other.

Chemical name
This is a written description of the chemical which is often complex. Not infrequently several such names may be used for one compound and consequently they are unsuitable for general use.

Common name
This term is used for general parlance when the chemical enters commercial use.

Trade name
This is the manufacturer's specific name for the active chemical material together with other substances that are added to facilitate easier application. Prior to the use of a trade name materials may be known by codes.

It is important to distinguish between active ingredient (a.i.) which is the active chemical present in material sold under a trade name and product which is a trade-named mixture for which the manufacturer provides application rates. For example, using benomyl as an illustration

molecular formula $C_{14}H_{18}N_4O_3$

structural formula

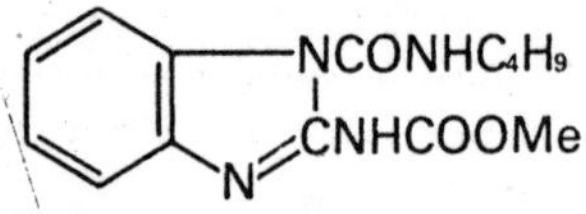

where Me = methyl radical

Chemical Name : Methyl 1-(butylcarbamoyl)benzimadozol-2-yl carbamate
Common Name : Benomyl
Code Number : Du Pont 1991
Trade Name : Benlate (a wettable powder containing 50 per cent active ingredient)

6.3 THE EVOLUTION OF FUNGICIDES AND BACTERICIDES

An evolution of chemical treatments can be traced from the use of inorganic materials via relatively unspecific organic compounds to the development of specific organic materials with some systemic properties (McCallan, 1967; Wain and Carter, 1977).

6.3.1 Inorganic materials

The value of elemental sulphur for control of rusts and moulds was known to Greek and Roman farmers. In the eighteenth and nineteenth centuries lime sulphur (sulphur boiled with unslaked lime) was used to control some powdery mildews of fruit trees and sulphur dust was used to combat powdery mildew of grapevine (*Uncinula necator*). Sulphur fungicides are now applied as dusts, where fungitoxicity is associated with particle size, or as wettable powders, slurries or paste formulations. These often contain wetting and dispersing agents to maintain suspendibility in store and during application. Other preparations include compounds with sodium, potassium, ammonium and polyethylene polysulphide. Sulphur-based fungicides are particularly effective against powdery mildew pathogens.

Use of copper fungicides began with the discovery of Bordeaux mixture (3 parts copper sulphate : 1 part calcium oxide : 100 parts water) for control of downy mildew of grapevine (*Plasmopara viticola*) by Millardet and Gayon (1885). Subsequent research led to refinements of this fungicide which have had widespread use. It is still employed particularly in underdeveloped areas because of low cost despite attendant risks from phytotoxic effects. Reaction of copper sulphate with other alkalis gave rise to Burgundy mixture (sodium carbonate), Azurin (ammonium hydroxide) and Cheshunt Compound (ammonium carbonate). The most effective formulation of Bordeaux mixture is achieved by adding concentrated lime paste to dilute copper sulphate solution. This has two drawbacks: (1) *the need for two separate containers*, one of which must withstand the corrosive effects of copper sulphate; and (2) *application is at very high volume*, which demands large volumes of water in order to apply the active ingredient at a concentration sufficiently low so as to avoid phytotoxic effects.

In the 1930s, mixtures containing fixed copper based on insoluble basic salts such as chlorides, sulphates, phosphates and silicates of copper and cuprous oxide began to be used. These could be formulated as wettable powders and did not require 'on farm' preparation. They also had the advantage of being less corrosive thereby dispensing with the need for special containers and they could be applied at higher concentrations which reduced the volume of spray to be applied. These materials were universally used for control of downy mildews and some other foliar pathogens. For specific purposes other copper compounds such as copper naphthenate, copper oximate and copper-zinc complexes have been utilised.

6.3.2 Non-specific organic materials

Both inorganic and organic compounds of mercury have been used for pathogen control and might be viewed as a bridge between these two forms of chemical. Widely used were mercurous chloride (calomel) and mercuric chloride (corrosive sublimate). The latter in weak aqueous solution was used to disinfect bulbs, corms and seeds for control of bacterial pathogens. Calomel slurry has had general application for control of *Plasmodiophora brassicae* (clubroot–section 2.6.1) affecting brassica transplanting material and to combat a wide range of turf pathogens. Extreme phytotoxicity limited the widespread use of inorganic mercury compounds but this was reduced by preparing organic formulations such as phenyl mercury acetate, phenyl mercury chloride, ethyl mercury chloride, ethyl mercury phosphate and ethoxy derivatives which were used as seed dressings. Their extreme mammalian toxicity has led to efforts to replace them with safer compounds. Development of organic fungicides gained momentum in the 1930s with the production of derivatives of dithiocarbamic acid. The first to achieve prominence was tetramethyl thiuram disulphide (thiram) which has had widespread use as a seed treatment. Thiram was followed by an iron derivative-ferric dimethyl dithiocarbamate (ferbam) used as a spray to control orchard pathogens such as apple cedar rust (*Gymnosporangium juniperi-virginianae*) and anthracnose (*Neofabraea malicorticis*). Subsequently a zinc salt-zinc dimethyl dithiocarbamate (ziram) was produced for control of early blight of potato and tomato (*Alternaria solani*) and tomato anthracnose (*Glomerella phomoides*). Quinone derivatives especially chloranil (tetrachloro-1,4 benzoquine) followed; this was extensively employed as a pea seed treatment. The range of dithiocarbamate materials was extended in the 1940s particularly with the development of a tank mix of zinc sulphate and disodium ethylene bis dithiocarbamate (dithane or nabam). The latter is unstable in air but is stabilised by zinc sulphate and widely used to control potato pathogens. Development of a stable form, obviating the need for tank mixing, led to zinc ethylene bis dithiocarbamate (zineb). Finally came the most effective of this group–manganese ethylene bis dithiocarbamate (maneb). Further details of this spectacularly useful group of fungicides are given by Thorn and Ludwig (1962).

Following the dithiocarbamates came carboximide derivatives particularly N-trichloromethyl mercapto-4-cyclohexImIde-1,2-dicarboximide (captan). This is active against a wide range of fungi affecting fruit, ornamentals and vegetables and is used for foliage, seed and soil treatments. It was followed by N-(trichloromethyl thio) phthalimide (folpet) used against species of *Alternaria*, *Botrytis*, *Pythium*, *Thanatephorus* and *Venturia* together with leaf-spotting, downy and powdery mildew pathogens infecting a wide range of crops. Other unrelated compounds were developed such as imidazolines (or glyoxalidines) which are heterocyclic nitrogenous materials, for example 2-heptadecyl-2-imidazoline (glyodin), and the guanidines such as n-dodecylguanidine acetate (dodine) both of which are active against *Venturia inaequalis* (apple scab—section 2.6.5). Also developed was dinitrocaprylphenyl crotonate (dinocap) which is effective against powdery mildew pathogens.

6.3.3 Specific organic materials

It was commented by Evans (1968) that: 'any future major contribution from the chemical industry (towards pathogen control) must in some way overcome the limitation of speed of epidemic development and accessibility of the pathogen. The problems of rapidly moving airborne epidemics would most likely be solved by the discovery of systemic materials capable of influencing the host/parasite balance at some point other than that of spore germination (for example during sporulation or growth)'. This statement emphasised developments that began before 1965 and have during the last 20 years provided a range of systemic compounds that possess very specific modes of action when compared with the rather non-specific actions of materials discussed in sections 6.3.1 and 6.3.2. Some antibiotics, thiabendazole and derivatives of organo-phosphorus insecticides such as ditalimfos and pyrazophos, were reported to have systemic properties prior to 1965 but the expansion in development and use of systemics followed the discovery of benomyl and carboxin in the late 1960s (Edington *et al*, 1980). A range of systemic materials is described in table 6.1.

Since then two horticulturally important but structurally unrelated non-systemic compounds have been developed: oxazolidine (vinclozolin) (Hess and Löcher, 1975) and hydantoin (iprodione) (Lacroix *et al.*, 1974). Vinclozolin is particularly effective against *Botrytis cinerea* (grey mould—section 2.6.4), *Monilia* spp. and *Sclerotinia sclerotiorum*-induced fruit rots. Iprodione has a broader spectrum of activity. Although non-systemic, they are absorbed into plant material and then degraded. They are especially useful as components of spray programmes enabling growers to use a range of chemicals and thereby to avoid a build-up of chemically tolerant pathogen strains.

Before the advent of synthetic systemic materials, considerable research was carried out into the potentialities of antibiotics and their derivatives. Although many of these substances were absorbed and translocated in higher plants and possessed antifungal and/or antibacterial properties they could not be used because of phytotoxicity problems. A few, such as the following, have come close to or reached commercial practice.

Table 6.1

Range of systemic fungicides

Groups	*Types*	*Antifungal activity*	*Reference*
Oxathiins (carboxylic acid anilides)	Carboxin Oxycarboxin Fenfuram Pyracarbolid Benodanil	High biological activity; particularly selective in effect against Basidiomycotina	von Schmeling and Kulka (1966)
Benzimidazoles Thiophanates	Carbendazim Benomyl Thiabendazole Thiophanate methyl	Wide spectrum of activity except against Mastigomycotina and Zygomycotina	Delp and Klopping (1968)
Pyrimidines	Ethirimol Dimethirimol Bupirimate	Particular action against powdery mildews	Elias *et al*. (1968); Finney *et al*. (1975)
Morpholines	Tridemorph Dodemorph	Particular action against powdery mildew	Kradel *et al*. (1969); Kradel and Pommer (1967)
Piperazines	Triforine	Wide spectrum of action against apple scab, powdery mildew, some rusts and leaf-spot pathogens	Schicke and Veen (1969)
Triazoles	Fluotrimazole	Foliar action against powdery mildews	Grewe and Büchel (1973)
	Triademefon	Active against powdery mildew, rusts and leaf-spot pathogens	Frohberger (1973)
Acylalanines	Metalaxyl	Particularly active against Oomvcetes	Urech *et al*. (1977)

1. *Streptomycin* This possesses broad activity against a range of bacterial species, for example, *Erwinia, Pseudomonas* and *Xanthomonas*. In particular it has been used in the USA against *Erwinia amylovora* (fire-blight); it does not cause russetting and gives more effective control than traditional Bordeaux mixture. There have been attempts to use dust formulations (section 6.4.2) but these failed due to lack of adhesion. Streptomycin has also been used against fungi such as *Phytophthora infestans* (late blight of potato and tomato—section 2.6.2) and *Pseudoperonospora humuli* (hop downy mildew). It possesses useful properties as a seed disinfectant for control of *Pseudomonas phaseolicola* (halo blight of *Phaseolus* beans—section 4.5) and *Corynebacterium betae* (silvering of beet). A tetracycline antibiotic, oxytetracycline, is added to streptomycin formulations to retard the rate at which antibiotic-resistant strains of *Erwinia amylovora* develop.
2. *Cycloheximide* This is active against fungi but not against bacteria; the use of cycloheximide has been inhibited by phytotoxic effects. For a time it was thought to have a valuable role against *Cronartium ribicola* (blister rust of *Pinus* and *Ribes*—section 2.6.8) but ultimately results were variable. It has been used commercially against some turf pathogens.
3. *Griseofulvin* This material possesses fungistatic properties and is active against a wide range of fungi except members of the Mastigomycotina; for example, various powdery mildews, *Botrytis* spp. and *Uromyces appendiculatus* (rust of *Phaseolus* bean) (Dekker, 1963; Wain and Carter, 1977).

6.4 MODES OF USE

6.4.1 Liquid sprays

The original chemical sprays were applied as more-or-less pure compounds or, as with Bordeaux mixture, were prepared from primary ingredients 'on site'. This practice has largely disappeared and most forms of crop-protection chemical are purchased as sophisticated formulations. For application these need only to be diluted with water and applied through some type of sprayer (chapter 7). The spray formulation as purchased will consist of at least some of the following.

Active ingredient
This is the active component of any formulated product.

Wetting agents
These compounds are incorporated into wettable powders to improve wetting or solution in the spray tank, and to solution and emulsion formulations to provide good penetration of pathogen lesions. They decrease

the contact angle of a spray droplet on the plant surface by reducing its surface tension. Wetting agents also tend to shorten flight and to curtail deposition of material from spray droplets on the plant surface (Dimond, 1962). Additionally the presence of a wetter in the dry residue hastens rewetting and attendant erosion and decomposition of the toxic agent. There are three types of wetting agent.

Anionic—consisting of positively charged particles

Cationic—consisting of negatively charged particles; if these are mixed, a precipitate forms that will clog up the sprayer

Non-ionic—consisting of uncharged particles; this type may be mixed with any other form of wetter and is the type most frequently used in agricultural sprays.

Spreading or sticking agents

In contrast to the small surface-active molecules of wetting agents these are of large molecular size although similarly surface-active; examples are proteinaceous materials, polymeric lattices or highly viscous oils (Somers, 1967). They aid the spread of spray droplets by lowering surface tension in the spray liquid. Additionally they increase the tenacity of spray deposits by forming cohesive gels over the dried deposit (Martin, 1964). This helps to resist weathering during short wet periods because such materials are only slightly soluble in cold water.

Dispersing agents

Composed of water-soluble cellulose derivatives these compounds are used to keep the fine particles of an insoluble fungicide in suspension.

Special additives

Materials such as trimethyl amine stearate are used to prevent the evaporation of fine aqueous droplets when sprays are applied from aircraft.

Safeners and adjuvants

Various compounds are incorporated into formulations to stabilise the active ingredients while in storage, to prevent decomposition of spray residues on the target plant or to neutralise phytotoxic decomposition products. Examples are urea, calcium carbonate and magnesium oxide. Materials such as glycerol are added to enhance uptake of an active ingredient possibly by retarding drying of the spray residue (Rich, 1956).

Fillers

Safe use of chemicals requires great dilution in application; the amount of active ingredient required at the leaf surface is 5–50 $\mu g\ cm^{-2}$ (Lukens, 1971), which is achieved partly by use of water as a spray diluent. Because only minute quantities of active ingredient need to be handled by the grower or spray contractor, preparation of a spray is made easier by selling active ingredients in forms that are already diluted by the addition of inert fillers such as china clay.

For spray application, the formulated product must be capable of even distribution in water. This is most easily achieved if the product is soluble in water. Most active ingredients are, however, insoluble in water and means of solving this problem are shown in table 6.2.

Table 6.2

Some of the more common types of spray formulation (after Evans, 1968)

Physico/chemical properties of an active ingredient	*Possible formulations*	*Appearance of spray dilution*
1. Substantially soluble in water	1.1 Granular or powdered solids that dissolve in water 1.2 Stock solution either in water or in a water-soluble solvent	Dilute solution of active ingredient
2. Substantially soluble in oil but not in water	2.1 Emulsifiable concentrate (e.c.) (miscible oil) active ingredient dissolved in oil 2.2 Stock emulsion of active ingredient dissolved in oil and emulsified with water	Dilute emulsion of active ingredient
3. Not substantially soluble in oil or water	3.1 Wettable powder (w.p.) freely dispersible in water 3.2 Suspension concentrate, a slurry composed of ground active ingredient suspended in water	Dilute suspension of active ingredient

When mixing different products in order to combat several pathogens or a pathogen and an insect pest in one spraying operation, it is essential that there should be no unwanted interaction between the components. This means that they should be *compatible*; the product label and manufacturer's literature should provide the necessary information. Mixtures of products should be used only where there is a positive recommendation to do so. Unwanted reactions that may totally change their properties can occur between components of different products.

6.4.2 Dusts, granules and seed dressings

'Dust' is a term used to describe fine dry particles of generally less than 30 μm diameter. These formulations contain 0.5 to 10 per cent of active ingredient although a few such as sulphur are applied without dilution. They are manufactured either by impregnating or coating highly adsorptive particles with an active chemical in solution or by mixing the active chemical and a filler together in a suitable mill. Typical fillers are fuller's earth, montmorillonite or kaolinite (Watkins and Norton, 1955); diatomite, pumice and talc are also used. Dust formulations are very useful in areas where water is scarce, but those with particle sizes of less than 30 μm constitute a health hazard to users. Also, such small particles are likely to drift and heavy rain will wash them from treated foliage. Equipment used to apply dusts usually consists of a hopper, preferably fitted with an agitator, a metering device that feeds the particles to a discharge outlet at a constant rate and a blower unit that produces an airstream to convey the particles towards a target.

Many of the disadvantages of dusts are overcome if the active ingredient is formulated into granules. These may be applied to row crops in a similar way to fertilisers. Granules are prepared by dissolving the active chemical in solvent and then impregnating a carrier, similar to those used for dusts. Large particles are prepared mechanically rather than grinding to a fine powder (table 6.3). The rate of release of active chemical will depend on the properties of the chemical, solvent and carrier. The period of effectiveness with granules is longer than that obtained with a single spray application. Granules can be treated with resinous plastics to control the rate of release and slow release into the soil may avoid loss of activity by adsorption on to soil particles. These types of formulation are used for fertilisers, insecticides and nematicides. Comparative size ranges of dust and granule particles are shown in table 6.3.

Dusts are widely used as a formulation for dressing seeds to prevent or reduce losses due to bacteria and fungi. There are two modes of action: (1)

Table 6.3

Size ranges of dust and granule particles (after Matthews, 1979)

Size range (μm)	*Description*
< 50	fine dust
50–100	coarse dust
100–200	microgranule
200–2000	granule
> 2000	macrogranule

destruction of pathogens carried with the seed either externally or internally, and (2) *protection of the seed* from attack by soil-borne pathogens during germination.

The advent of systemic fungicides has permitted the control of pathogens situated within seeds and previously inaccessible to chemicals. They were mostly controlled by hot-water treatment; this was a hazardous process which if not propertly monitored could kill the seed. Also, sowing had to follow treatment quickly or the seed needed to be redried. Systemic dressings also provide an opportunity to treat seed with a reservoir of chemicals that will be active against pathogens during crop growth. Where such prolonged systemic effect is required, large quantities of active material must be applied to the seed to overcome the dilution effect of crop growth and breakdown of active ingredients in soil. Quantities of chemical that may be applied to seed are restricted by the physical nature of an active compound, type of dressing machinery used and retentive capacity of the seed. When treated seed is sown, some dressing is inevitably lost but sufficient needs to remain on the seed and in its close proximity in order to enter the developing root system. The success of seed dressing as a technique for control of pathogens is shown by the widespread use that has been made of thiram and, more recently, by rapid utilisation of benomyl and iprodione especially for control of *Botrytis* spp. causing onion neck rot and *Sclerotium cepivorum* (onion white rot).

Seed dressing is an economical and convenient method of control; quantities of chemical required are small and the dressing operation is easily mechanised. It may be adapted to factory operation requiring limited specialised equipment. Treated seeds leave minimal residues in soil or harvested crop. It is necessary to ensure that the active ingredient is evenly distributed over a given quantity of seed to produce maximum biological effect. Formulations employed are oil-based dusts (1 per cent oil providing adhesion), concentrated slurries and concentrated solutions of active ingredient. These are applied to seed by methods such as those shown in table 6.4.

Most seed dressings are applied as adhesive dusts. This avoids the complications that arise from increasing seed moisture content. In batch

Table 6.4

Summary of seed-dressing techniques (after Evans, 1977)

Formulation	*Application technique*
Adhesive dust	Batch or continuous flow process
Wet slurry Concentrated solution	Dribble or spray application

dressing (Jacks, 1953; Sharvelle, 1961) a standard quantity of seed and chemical are placed in a drum which is rotated until an adequate distribution of dressing has been achieved. Providing manufacturers' recommendations are followed, dust formulations will distribute uniformly over the seeds, leaving only minimal quantities on the container wall. In continuous-flow machinery, mixing is achieved in a suitable tumbling device into which metered quantities of seed and dressing are admitted and the treated seed continuously removed by a worm screw for bagging. Slurry or concentrated-solution preparations remove the hazards of dust application for operators but add 0.5–1 per cent moisture to the seed. They may be mixed with seeds in a mixing tank, or from a more recent development, be sprayed on to a continuous flow of seeds.

Overdosing seeds with dressing may result in malformed seedlings or in a failure to germinate; these effects are accentuated by previous injury to the testa or by storage at high temperatures. The condition of treated seed is of prime importance in determining whether or not injury will occur. To avoid phytotoxicity, seed should be undamaged, stored at low temperatures, of low moisture content and the chemical should be used at manufacturers' recommended rates.

6.4.3 Soil fumigants

Where sterilisation is still practised for protected crops, chemical fumigants have largely replaced the use of steam (section 9.1.2). Chemical fumigants (volatile chemicals) have been used on a field scale for the past 30 years in the USA to control pathogens such as *Verticillium* spp. on areas monocropped with strawberries and tomatoes. This method is becoming increasingly popular in Europe particularly for high-value nursery-stock land, soft-fruit plantations with restricted rotations and top-fruit orchards where replant disorders occur. In order of activity the fastest and furthest reaching fumigants are methyl bromide, chloropicrin (trichloronitromethane), 1, 3-D (1,3-dichloropropene), 1,2-dibromoethane, methyl isothiocyanate and 1,2-dibromo-3-chloropropane. Most fumigants are either low boiling-point liquids or solids which decompose rapidly forming volatile compounds. Consequently it is essential to seal the soil surface for a period after application by compaction, flooding to a depth of 1–2 cm or use of polyethylene sheets. Chloropicrin, DD-mixture (dichloropropene-dichloropropane) and methyl isothiocyanate are examples of low boiling-point liquids which can be injected into the soil. Methyl bromide gas is usually injected beneath polyethylene sheeting and allowed to diffuse into the soil surface layers. Only methyl bromide, chloropicrin and methyl isothiocyanate are fungicidal at field-dose rates (Barnes *et al.*, 1966). Others such as dibromoethane are largely nematicidal and may be used to control the vectors of soil-borne viruses.

Under glass, gaseous or low boiling-point liquids may be injected from single or multipoint applicators whereby metered amounts of chemical are forced into the soil under pressure. In the field, single line or multi-line injection is achieved using chisel plough bodies into which are incorporated tubes feeding fumigant from a reservoir to the plough bottom under gravity. A disadvantage with this technique is that full treatment rates are applied only to the ploughed strips and the remaining soil receives lower doses by diffusion. This problem is overcome by use of single-plane applicators which are of two types.

1. *Blade applicator*—in which fan-type spray nozzles (section 7.1.4) are fitted beneath a blade in a recessed manner. This is drawn through the soil at a controlled depth, fumigant passes from a reservoir and is pumped out through the nozzles, providing a continuous band of treated soil.
2. *Sweep applicator*—which functions in a similar manner to the blade one but consists of several separate triangular sweeps each fitted with a fan nozzle 7–8 cm behind the sweep point. Correct spacing of the sweeps on a tool bar permits continuous treatment of a band of soil. The advantage of sweeps is that they clear themselves of debris which tends to build up in front of single-blade applicators (Page, 1963).

With either tool, treated land must be sealed to retain the fumigant. This is best achieved by a separately powered roller that seals but does not compress or shear the soil surface. Where dust or granular fumigant formulations are used they are either left undisturbed on the surface or, more commonly, worked into the upper soil horizon by rotary tillage or discing. Dazomet, which in soil breaks down to yield methyl isothiocyanate gas at temperatures above 7°C, is formulated in a solid prill or granular state which is mixed into soil by rotary cultivation.

Factors that influence movement of fumigants are the chemical and adsorptive properties of the toxicant, and the temperature, moisture, organic matter, texture and profile variability of the soil. Fumigants applied as liquids or solids first vapourise to a gaseous phase. Individual molecules then have an affinity for and may be removed by water, mineral and organic phases within the soil. Fumigant diffusion is greatest towards the atmosphere rather than towards the remaining soil. Movement by fumigants in soil is determined by their vapour pressure characteristics, that is, the rate of change from liquid or solid state to gaseous state. For those with a high vapour pressure, for example methyl bromide, concentration will increase rapidly in the soil pore spaces, being initially influenced by mass flow and then by diffusion. For those with low vapour pressure characteristics, for example ethylene dibromide, there is slow movement into the vapour phase with little mass flow except in very warm soils or where water is added following fumigation. Fumigant molecules dissolve in soil water to establish a concentration equilibrium between soil water and soil air. This equilibrium obeys Henry's Law which states that the solubility of a gas in a solvent, with which it does not react chemically, is directly proportional to the pressure of the gas. Consequently the concentration of fumigant in soil air:soil water should be constant at specified temperatures but this is modified by the complex nature of soil mixtures. Factors such as the effects of water:air and water:solid interfaces, the concentrations of other soil ingredients and the nature of the fumigant will affect this equilibrium. The effects of solubility in soil water influence the ability of fumigants to destroy those soil-borne organisms which are themselves bathed in a water film and otherwise affords protection from fumigants in a gaseous state. In moist soils only limited amounts of fumigant are adsorbed on to soil particles. Where soils are very wet and the passageways between particles are blocked, fumigant movement is decreased and concentration is diluted by excessive solution in soil water. The rate of fumigant diffusion through water is negligible in comparison with diffusion through air. Consequently, fumigation of wet soils or fumigation followed by heavy rain is of limited value. In general, the rate of diffusion of fumigants through soil will be influenced by the molecular weight, temperature, presence of co-diffusing gases, continuity of air spaces, and distribution of the fumigant between air, water and solid phases within the soil. The latter is also affected by soil texture and in general coarse-textured soils fumigate more successfully than those of a fine texture such as clay loams and clay. Mineral content has little effect on fumigation but the percentage of organic matter on to which a fumigant may adsorb is very important. Soils with a high organic-matter content adsorb fumigants more strongly than lighter sandy soils and retain them longer so that greater intervals are required between treatment and sowing or planting. To achieve successful fumigation, land must be thoroughly worked before treatment and should be evenly moist but not wet because adequate distribution of the fumigant is essential. Soils that cannot be worked to a reasonable tilth or contain large stones are unsuitable for treatment since clods of soil are not penetrated by the chemical and stones prevent adequate sealing of the soil surface after treatment.

The fumigant ultimately decomposes or diffuses until the concentration in the soil becomes negligible. Following fumigation at rates sufficient to destroy fungi and nematodes, there is a period of sterilisation or partial sterilisation of the treated soil. This releases nutrients from the dead biomass which are available to any surviving or reinvading flora. Crop plants generally produce superior growth in treated soil (Wilhelm and Nelson, 1970). Fumigation eliminates or reduces the obvious and less-obvious soil pathogens and pests, reduces weed competition and alters soil nitrogen content. Fumigated soils tend to contain more ammonium nitrogen. In cool damp soil this may cause poorer plant growth, which might also be attributed to an elimination of mycorrhizal associations or presence of toxic residues. Freedom from such residues can be tested by ensuring that rape seed will germinate normally on the moistened surface of a 1 kg soil sample placed in a gas-tight glass container.

Materials such as formaldehyde, some formulations of methyl isothiocyanate and phenolic compounds may be applied by a grower provided that adequate safety precautions are observed. Other compounds should be applied only by contractors who are trained and equipped to deal with dangerous chemicals. The use of methyl bromide, in particular, is covered by special regulations designed to minimise the risks inherent in its use.

6.4.4 Smokes and fogs

Some chemicals, particularly for protected crops, are applied as smokes or fogs but this can be done only where the active ingredient is heat stable. Smokes are formulated as part of a pyrotechnic mixture which ensures vapourisation but not oxidation. Generators for smokes are carefully designed to avoid explosion and permit a controlled rate of burning. The high velocity of hot gas emitted causes mixing of an active ingredient with the surrounding atmosphere before condensation produces a fine smoke. High temperatures must exist for short periods only in order to minimise breakdown of an active ingredient. This method of application provides excellent penetration of crop canopies with uniform deposition but relatively brief persistence. For such treatments, plant foliage should be dry and the soil moist. The method should not be used on seedlings, newly planted crops or in bright sunlight otherwise phytotoxic damage may occur.

In fogging techniques the active ingredient is dissolved in oil (such as paraffin or diesel oil) with a flash point above 65°C, possibly with an additional co-solvent to prevent sludge formation. A fog may be defined as when aerosol droplets of less than 15 μm diameter fill an air volume to such an extent that visibility is reduced. In Europe this technique has been used for protected crops, elsewhere there are some field applications. Environmental restrictions to avoid phytotoxic damage are similar to those for smokes. Care must be taken to avoid inhalation of a fog preparation because the smaller droplets are not trapped by nasal membranes but will carry into the lungs. Protective clothing with a full-face respirator is essential.

Fogs are generated by injecting pre-heated chemicals into the exhaust systems of engines burning petrol or liquid petroleum gas or, with fan assistance, over electrically heated coils where hot air vapourises the chemicals which are impregnated on a cartridge. Where combustion engines are used the exhaust gases are pressurised by passage at high velocity along a pipe of smaller diameter than the combustion chamber. Active chemical may be admitted either at the end of this tube or close to the combustion chamber from a pressurised reservoir tank.

6.5 MODES OF FUNGICIDE AND BACTERICIDE ACTION

For successful action against a pathogen the chemical needs to be present at the 'correct' site at the 'correct' time and in sufficient quantities to be lethal. Traditionally this might be taken to mean that soil-borne pathogens are combated with soil-applied materials, seed-borne pathogens by seed dressings and airborne pathogens by foliar application. Essentially this requires that a pathogen is attacked at the weakest point in its life cycle, that is, when the inoculum potential is smallest (Garrett, 1956). This might be equated with conditions being unsuitable for infection, or when the vigour of inoculum is lowest or a host is least susceptible. At best this means the dormant or pre-planting phase for seed treatments or the pre-cropping phase for soil fumigants. On this basis chemicals could also be applied at stages in a host life cycle when it is least likely to be affected by the phytotoxic effects of large amounts of protectant chemical. An example is the application of organo mercurial compounds to apples following harvest in one season and pre-bud burst in the next to control *Venturia inaequalis* (apple scab—section 2.6.5); this minimises chemical residues in the fruit and phytotoxic hazards to the tree (Burchill and Hutton, 1965).

Fungicides exert their action by at least three pathways (van der Kerk, 1971).

1. *Inhibition of energy production* They affect ATP (adenosine triphosphate) production by, for example, inhibition of respiration or by uncoupling oxidative phosphorylation. These energy-producing processes are located partly in the cytoplasm and partly in the mitochondria. Strong inhibition of such processes will eventually have a lethal effect.
2. *Interference with biosynthesis.* This is disruption of those processes leading to production of new cell material and maintenance of the pathogen organism. Biosynthesis of low molecular weight compounds such as amino acids, purines, pyrimidines and vitamins takes place in the cytoplasm while protein synthesis proceeds mainly on the ribosomes and DNA and part of RNA synthesis in the nucleus. Inhibition of biosynthesis may only have a fungistatic effect especially where chemicals are at low concentration.
3. *Interference with cell structure* For instance, by affecting the permeability of the cell membranes, leakage of the cell contents will result and this may have fungicidal effects.

Most non-systemic materials such as dialkyl dithiocarbamates, bis dithiocarbamates, captan, dinocap, triphenyltin acetate, chlorothalonil and phenyl mercury acetate act by inhibition of energy-producing processes. Some systemic compounds affect biosynthetic processes; others inhibit respiration. The modes of action of some fungicides are shown in table 6.5.

Table 6.5

Mode of action of some fungicides (after Kaars Sijpesteijn and van der Kerk, 1970; Dekker, 1977)

Compound	*Inhibitory action based on interference with*
Benomyl Thiophanate methyl Thiabendazole Fuberidazole	DNA biosynthesis
Chloroneb	DNA biosynthesis
Carboxin	Respiration, biosynthesis of DNA and RNA
Dimethirimol Ethirimol	Folic acid-mediated C_1 transfer reactions
Azauracil	Biosynthesis of UMP
Sulphanilamide	Biosynthesis of folic acid
Griseofulvin	Cell division
Cycloheximide	Protein biosynthesis
Polyoxin B	Chitin biosynthesis
Pimaricin Dodine	Disruption of cell membranes

Protectant and non-systemic eradicants are mainly general cell poisons to both host and pathogen. They are selective in effect against the parasite because they do not penetrate beyond the host plant surface. The cuticle acts as a barrier preventing damage to cells beneath. When these materials penetrate below the cuticle, damage ensues since there is little margin of safety between fungitoxicity and phytotoxicity. Systemic materials must be able to enter a host and be toxic to the pathogen without killing the plant. As the pathogen grows more quickly than its host it is more vulnerable to chemicals that inhibit biosynthetic pathways. In addition, growth in higher plants is limited to meristematic tissues.

Non-systemic materials are applied in advance of infection to form a barrier to invasion; hence it is necessary to anticipate infection and further applications are required as new tissue is produced. Systemic materials should improve on this situation in the following ways.

1. By eradicating existing infections.
2. By providing permanent suppression of sporulation thereby inhibiting epidemic development.
3. They are not subject to processes of weathering, limitations in method of application or variations in plant surface.
4. They may be applied after infection has taken place and can eliminate costly wastage inherent in incorrectly timed spray programmes where the onset of an epidemic is not predictable with any degree of certainty.

Systemic action avoids the need to obtain saturation spray cover as was demanded by earlier fungicides and hence permits low-volume application (section 7.2.4) with consequent savings in energy and water.

Characteristics of current systemic fungicides are as follows.

1. They possess direct fungitoxicity and control pathogens by modifying the host.
2. They have low to very low water solubilities.
3. They enter the aerial tissues and plant roots and pass into the xylem tissues.
4. They move upwards in the transpiration stream and may accumulate at leaf margins.
5. They do not enter organs that are incapable of transpiration nor are they re-exported to regions of new growth.
6. Downward movement, if any, in the phloem has no practical significance.
7. Activity spectra are either very wide or very restricted (Kirby, 1972; Sbragia, 1975).

The obstacles that a systemic chemical must overcome are as follows.

1. Host cuticle and subcuticular cells.
2. Metabolism in the host.
3. Adsorption in the host during translocation.
4. Metabolism by the pathogen.
5. The pathogen membrane(s) (Bryde, 1969).

The degree of availability and movement of systemic fungicides are highly dependent on their chemical and physical properties, the nature and growth stage of the host, the form of translocation in the host, the method of application and environmental factors (Evans, 1971).

Systemic chemicals may modify and be modified by the host plant. Some possess *in vivo* activity which may not be apparent *in vitro*. In such cases it is breakdown products of the chemical in plant tissue which are fungicidal. Other materials are fungitoxic both *in vivo* and *in vitro*. To describe these situations the active chemical moiety within a host is termed *toxophore* and a chemical that enters a host but is foreign to it is termed *xenobiotic* (this will also include insecticides and herbicides). Xenobiotics first enter into the free space within tissues, that is, those parts of plant tissue that communicate freely, by diffusion, with the external environment (Kramer, 1957; Biggs and Robertson, 1957). Then they move in either of the following ways.

1. In the direction of the evapo-transpiration stream; this is *apoplastic movement.* The apoplast is the non-living part of a plant, such as cell walls, cuticle, xylem vessels and tracheids. Movement in these tissues is passive and does not require the expenditure of energy.
2. With photosynthates towards 'sinks'; this is *symplastic movement*. The symplast is the living part of a plant enclosed by membranes, for example, protoplasts, plasmadesmata and phloem sieve cells. This movement is active, requiring the expenditure of metabolic energy. Long-distance transport within the symplast system takes place via phloem cells, probably in the sieve tubes.

Symplastic and apoplastic parts of a plant are separated by the membrane which bounds the protoplast, that is, the *plasmalemma*. Some materials are capable of both apoplastic and symplastic movement and are termed *ambimobile* (Edington, 1981). Physiological requirements for systemic activity are as follows.

Membrane permeability

Most systemics move mainly upwards in plants via the apoplast but all must be capable of traversing some symplastic tissue. When entering via the root system they are required to cross the endodermis through a protoplast because of the Casparian strip. Possibly they also enter vacuoles and other membrane-bounded organelles such as mitochondria and chloroplasts.

Selective toxicity

Because systemic compounds enter the protoplasts they must be non-toxic to host tissue; they must not be rapidly degraded by it but should be active against a target pathogen. For example, carbendazim possesses selective affinity to the microtubulin protein of fungi but not to the microtubulin protein of higher plants (Davidse, 1973) while carboxin is selectively toxic to fungal succinic acid dehydrogenase enzyme complex but not to that of higher plants (Mathre, 1970).

Metabolic stability

Systemics entering the symplast are subjected to metabolic processes within host cells and consequently need to remain unaffected by such processes in order to reach a pathogen. Stability varies from one host species to another; for example, triadimefon applied as a soil drench will control fusiform rust (*Cronartium quercuum* f. sp. *fusiforme*) of Loblolly pine (*Pinus taeda*) for up to 40 days but degrades rapidly in other plants (Clark *et al.*, 1978; Gastonyi and Josepovits, 1979; Kelley, 1980).

Cuticular penetration

Systemics are principally applied as sprays; consequently a major barrier to penetration is the cuticle. Fungicides partition from diluted spray, diffuse through the cuticle and partition into the apoplast below initial droplets. Because spray droplets dry rapidly, systemic materials are formulated as emulsifiable concentrates or oil dispersions (section 6.4.1). These permit rapid liquid-to-liquid partitioning into water and thence into the cuticle. If a xenobiotic is lipophilic, penetration is fast; however, if it is too greatly lipophilic there may be retention by cuticular waxes and consequently retardation of transcuticular diffusion and subsequent release into the apoplast (Price, 1977). Very hydrophilic molecules will partition less rapidly into the cuticle but will continue to be partitioned from cuticle to apoplast for a prolonged period leading to greater total uptake (Edington *et al.*, 1978).

Water solubility

Water solubility does not control permeability through membranes; it is important, however, in delivering the systemic to a plant surface in a soluble state. Benomyl, for example, is less successful as a soil drench than thiophanate methyl because of its lower water solubility despite both materials being converted to MBC (methyl benzimidazole-2-yl-carbamate) within the host.

6.6 PATHOGEN TOLERANCE TO CHEMICALS

Some instances have been reported where tolerance to non-systemic chemicals has developed in pathogens; for example, tolerance by *Venturia inaequalis* (apple scab—section 2.6.5) to repeated use of dodine (Szkolnik and Gilpatrick, 1973; Gilpatrick and Blowers, 1974). Instances of tolerance have devloped very much more quickly with systemic materials. With both types of chemical, tolerant fungal strains have occurred predominantly as a result of selection rather than mutagenesis, although tolerance to PCNB (pentachloronitrobenzene) in *Hypomyces solani* f. sp. *cucurbitae* may be mutational in origin (Georgopoulos, 1962). One of the first instances of tolerance to systemic fungicides was reported with *Sphaerotheca fuliginea* (cucumber powdery mildew) and the pyrimidine derivative dimethirimol. This was introduced into the Netherlands in 1968 and used extensively. By 1970, reports of a loss of efficacy began to emerge and dimethirimol-tolerant strains of *S. fuliginea* identified (Bent *et al.*, 1971). Tolerance to systemic fungicides has been reported for species of *Botrytis*, *Cercospora*, *Colletotrichum*, *Erysiphe*, *Fusarium* and *Verticillium* (Dekker, 1977).

Development of tolerance under field conditions is influenced by the way in which fungicides are used. Application of high doses reduces the rate of emergence of tolerant strains. Once they have appeared, however, continuation of this policy maintains a high selection pressure, exacerbating the problem by promoting an expansion of tolerant populations. The method of chemical use may also be influential when tolerant strains have developed, especially after in-furrow or seed treatment. These techniques ensure that fungicides remain present at lethal or, more importantly, sub-lethal concentrations for an extended period (Wolfe and Dinoor, 1973). Where the emergence of tolerant strains can be anticipated, application should be of alternative chemicals in sequence or mixtures of chemicals—preferably each possessing a different mode of action. Integrated use of benomyl and dimethirimol was not followed by a development of *S. fuliginea* strains tolerant to both chemicals (Ebben and Spencer, 1973). A combination of systemic and non-systemic materials is useful since this reduces sporulation and subsequent spread of mutants.

6.6.1 Types of tolerance

Decreased permeability

Excessive use of blasticidin S, which inhibits spore germination and mycelial growth of *Pyricularia oryzae* (rice blast) through inhibition of protein synthesis, has led, for example, to a type of tolerance attributed to an inability by this antibiotic to reach sites of protein synthesis because of decreased permeability of protoplast membranes in tolerant pathogen isolates (Huang *et al.*, 1964).

Conversion to an inactive compound

Differences in an ability to convert PCNB (pentachloronitrobenzene) into less fungitoxic compounds appear to be a basis for variation in sensitivity among strains of *Botrytis cinerea* (grey mould—section 2.6.4) (Nakanishi and Oku, 1969). Development of tolerance to dodine by *Nectria haematococca* f. sp. *cucurbitae* strains following irradiation with ultraviolet light may be related to an increased capacity for converting dodine to less fungitoxic compounds (Kappas and Georgopoulos, 1970).

Lack of conversion into a fungitoxic compound

In a few cases, an initially inactive compound is converted by a pathogen into a toxic material (*lethal synthesis*); thus for fungitoxic effect, 6-azauracil (AzU) is converted via 6-azauridine (AzUR) into 6-azauridine-5′-phosphate (AzUMP). Only this latter compound is a competitive inhibitor in the target pathogen for orotidine-5′-phosphate decarboxylase (OMP decarboxylase), an enzyme that functions in pyrimidine biosynthesis (Dekker and Oort, 1964). Irradiation of *Cladosporium cucumerinum* (cucumber scab) with ultraviolet light, however, leads to strains tolerant to AzU because they are incapable of converting it to AzUR (Dekker, 1967).

Detoxication by binding

A failure of organo mercury compounds to control *Pyrenophora avenae* (stat. conid. *Drechslera avenae*) on oats is claimed to be due to production of red anthraquinone compounds by tolerant strains which bind to the organo mercurial compound and thereby inhibit their fungitoxic effect (Greenaway, 1971). This work has not, however, been repeated elsewhere.

Decreased affinity at the reactive site

Benomyl and thiophanate-methyl are converted to MBC (methyl benzimidazole-2-yl-carbamate) within a host and this is fungitoxic by virtue of inhibition of chromosome separation at mitosis. The close correlation of pathogen sensitivity to MBC and its binding to some macromolecules that have similar properties to microtubular protein is thought to be a means by which MBC prevents assembly of microtubular subunits into functional spindle fibres, thereby inhibiting mitosis. Development of tolerance is ascribed to a change in protein subunits of the pathogen spindle microtubules resulting in a decreased affinity to MBC (Davidse, 1975a, b).

Circumvention

There is some evidence that development of tolerance to carboxin by *Ustilago hordei* (covered smut of barley) is at least partially dependent on a development of alternative metabolic pathways such as the glyoxylate cycle or carboxin-tolerant phosphate uptake (Ben-Yephet *et al.*, 1975).

Compensation

Tolerance may develop when changes in the metabolism of a sensitive organism result in less of a product of an inhibited reaction being required or the quantity of an inhibited enzyme being increased. Tolerance of this type is indicated by a 6-azauracil-tolerant strain of *Cladosporium cucumerinum* (cucumber scab) which produced at least three times more of a target enzyme than a sensitive wild-type fungus (Dekker, 1971).

Other effects

Tolerance by strains of *Verticillium malthousei* (brown spot of mushroom) *in vitro* to benomyl and other unrelated chemicals was attributed to a capacity by these strains to acidify the agar medium which might influence the degree of uptake and binding of the fungicides (Lambert and Wuest, 1975).

Problems that have arisen with systemic chemicals include the following.

1. *Development of tolerant pathogen strains* possibly because of their highly specific mode of action in comparison with the general cell-killing properties of earlier non-systemic materials.
2. *Greater persistence in crop tissues* leading to longer periods over which residues are present.
3. *Alterations in the ecological balance* to such an extent that micro-organisms that are not sensitive to a particular chemical increase in significance as sensitive pathogens are controlled; for example, the increased importance of *Rhizopus stolonifer* in strawberry and *Penicillium* spp. on tulip bulbs as a result of the use of benomyl.

6.7 ECONOMICS OF CHEMICAL APPLICATION

Botrytis cinerea, the cause of grey mould (section 2.6.4) in strawberry, is an example of a pathogen that can devastate this crop at a very vulnerable growth stage. This pathogen causes rotting of both green and ripening fruits, quickly reducing them to soft masses covered with dusty grey fructifications. It is responsible also for reduction in yield because of the induced abortion of flower trusses. The greatest danger comes from early infections of floral organs or by contact with previously infected fruits. Infections established early in fruit formation may remain latent until sporulation and subsequent rotting develop at harvest. Losses of ripe fruit coincide with the stage when a grower has spent most of his money required to produce the crop and is about to obtain financial returns from sales. This pathogen is also an example of one where there are few means of control other than by the use of chemicals. Little or no resistance is available to control *B. cinerea* and the most widely used cv. Cambridge Favourite, which accounts for 70–80 per cent of the cropped area, is highly susceptible.

Husbandry techniques including the following may be used to enhance chemical control but they would be totally inadequate if the grower relied solely upon them.

1. *Strict regulation of the use of nitrogenous fertilisers* in fruiting plantations to minimise the production of lush foliage and fruit with excessively high water content, which are both thought to render a crop more susceptible to infection.
2. *Frequent picking* to remove all ripening fruit including undersized berries and overripe fruit which if left on the plant would form foci for infection by *B. cinerea*.
3. *General plantation hygiene* to remove or destroy latent sources of infection; this includes rotovation between the matted rows to destroy moribund tissue infected with *B. cinerea*. It is thought that use of dinoseb-in-oil used to burn off unwanted runners between matted rows will help to eliminate sources of *B. cinerea* infection.

The main means of control, however, is the use of fungicides. Sixteen products are listed for control of *B. cinerea* in UK strawberry plantations (Anon, 1982a). There are, however, circumstances whereby this list often becomes considerably shortened. Materials such as captan and thiram may not be used where fruits are destined for processing (a major market outlet) because there is a risk of tainting. Where fungicide-resistant strains of *B. cinerea* are identified (section 6.6), materials such as benomyl, carbendazim and thiophanate-methyl have no effect in controlling this pathogen. This has led to a situation where growers are strongly advised to use several fungicides in rotation. The materials available are dichlofluanid, iprodione, chlorothalonil and vinclozolin.

Effective control of *B. cinerea* is obtained by use of a thorough programme of fungicide application prior to and during blossoming and in the fruit-swelling periods. As a result of the long flowering period of strawberry it is advisable to apply sprays at the following times.

1. *At pre-flower emergence* to reduce any residual infection on old leaves and moribund fruits remaining from the previous season.
2. *At white bud or first open flower.*
3. *At early flowering* when the first flowers are open on a majority of plants.
4. *At full blossom* when a majority of flowers are open on most plants.
5. *At petal fall* when most petals have fallen on the majority of plants.

In practice, half of these sprays are applied at less than ideal growth stages because wind and rain reduce the number of days that are suitable for spraying. A suitable spray programme and its costs are shown in table 6.6.

From table 6.6 it can be seen that the total cost of a chemical spray programme for control of *B. cinerea* (grey mould) in strawberry is £111.78 ha^{-1}. Assuming an average yield of 10 tonnes ha^{-1} of fruit, the cost of

Table 6.6

*A spray programme for control of grey mould (*Botrytis cinerea*) of strawberry*

	Chemicals	*Price (1981) (£ ha^{-1})*
Stage 1	Dichlofluanid applied overall at 1.7 kg product ha^{-1} at £8.83 kg^{-1}	15.01
Stages 2, 3 and 4	Iprodione applied as a band spray, at 1 kg product ha^{-1} at £23.30 kg^{-1}; the matted row cultural system for strawberry produces bands up to 450 mm wide, hence 1 kg product will treat 2.2 ha field area and the cost per ha of 3 sprays will be $\frac{23.30}{2.2} \times 3$	31.77
Stage 5	Vinclozolin applied as a band spray (this material is used at this point in the programme because of the shorter interval allowed between application and picking) at 1.5 kg ha^{-1} at £22 kg^{-1}, cost will be $\frac{22.0}{2.2} \times 1.5$	15.0
	plus approximately £50 ha^{-1} for application costs assuming that a contractor is used	50.00
		111.78

spraying per kg of picked fruit is 111.78/10,000 = £0.011 kg^{-1}. With returns of, for example, £0.35 kg^{-1} for fruit destined for processing and £0.80 kg^{-1} for fresh market fruit, spraying costs 3 per cent and 1 per cent respectively of crop market value. Investment in a reliable spray programme to control *B. cinerea* is a sound economic proposition.

In economic contrast is the possibility for large-scale control of clubroot (*Plasmodiophora brassicae*—section 2.6.1) by means of soil sterilisation. Investigations have shown that basamid will significantly reduce the level of *P. brassicae* in soil but will not eliminate this pathogen. Basamid at 717 kg ha^{-1} reduced the severity of infection to 0.1 per cent of that found in controls (Brown and Melville, 1970) while, when applied at 400 kg ha^{-1}, the number of affected plants was about 10 per cent of that in untreated infected plots (Buczacki and White, 1979). At 1981 prices basamid cost £2.31 kg^{-1}; consequently treatment at these rates would cost £1656.27 ha^{-1} and £924 ha^{-1} respectively, neglecting the cost of application. These figures may be compared with returns for cabbage crops shown in table 6.7.

Table 6.7

Returns for cabbage crops (Anon, 1982b)

Cabbage crop	*Yield ($t\ ha^{-1}$)*	*Price (£ t^{-1})*	*Gross value (£)*
Summer	15	112	1680
Autumn	20	95	1900
Winter	25	110	2750
Spring	12	181	2172

Clearly, although basamid could theoretically be used for this purpose, in practice this might be contemplated only where costs could be distributed over a long term or where small areas of infection could be treated on valuable and otherwise pathogen-free land. Additionally since total eradication is not achieved, invasion would re-occur to detectable levels in the foreseeable future when susceptible hosts were reintroduced to the land. Also it would be impossible to prevent reinfestation from sources outside the treated area.

6.8 REGULATION AND APPROVAL OF CHEMICALS

Techniques for discovery of new chemicals are still largely empirical despite much basic work to determine the types of structure likely to be active against plant pathogens. Agrochemical companies employ chemists to synthesise novel compounds either as analogues of existing materials with known activity or as entirely new structures. These then pass through a biological screen in controlled conditions using small numbers of crop pathogens to show which materials have activity. Promising materials are passed through a wider screen using a greater number of pathogens. Materials with positive action are then tested in the field. At this point development work with the material begins. The agrochemical company must find the most efficient and effective formulation by which the chemical can be applied and also begin to collect data that will permit acceptance by government regulatory organisations concerned with safety in all its aspects.

If the criteria listed in section 6.1 were always fulfilled, clearance procedures would be unnecessary but in reality all chemicals offer a potential hazard to users, consumers and/or the environment. Consequently regulatory organisations have been established in Europe. North America, Australia and Japan to prevent the marketing of very hazardous materials. In the UK the safety of pesticides is covered by the *Pesticide*

Safety Precautions Scheme (PSPS) which is a formal but flexible voluntarily agreed scheme between Government and the agrochemical industry. Under this scheme manufacturers have undertaken not to market products containing any new chemical, not to introduce new uses for existing chemicals or new formulations until recommendations for safe use have been agreed with the government departments concerned. The processes that must be gone through to obtain clearance in the UK are shown in figures 6.1 and 6.2.

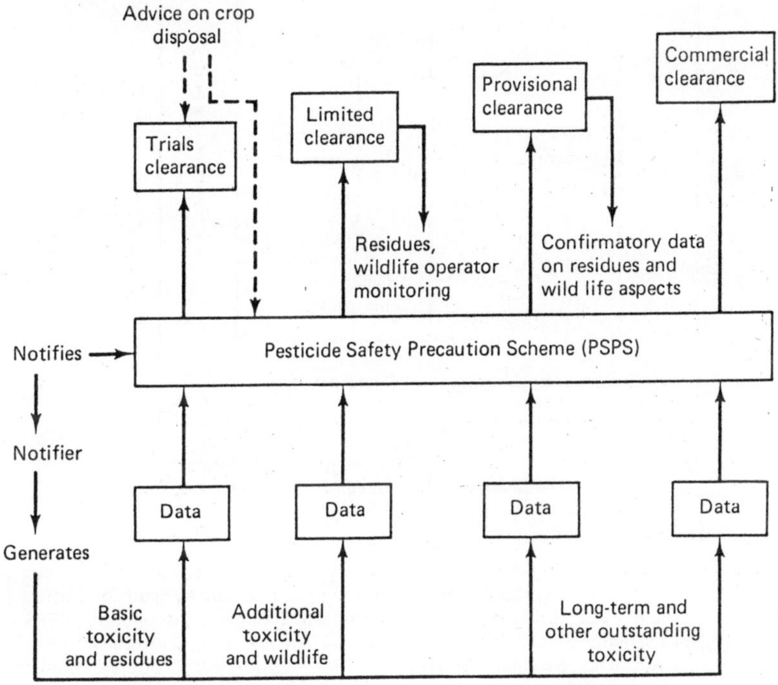

Figure 6.1 Illustrating the levels of clearance for use given to agricultural chemicals in the UK between first notification of the regulatory authority and full commercial clearance (reproduced by permission of Ministry of Agriculture, Fisheries and Food, London)

The chemical passes through four levels of clearance (figure 6.1) based on the accumulation of data concerning its physical, chemical and biological properties, and its persistence in crops and soil; products that may break down in the plant, their mode of action, their toxicity to mammals and, where available, to man, and their likely effects on wildlife including birds, fish and bees; the proposed use and methods of application and risks

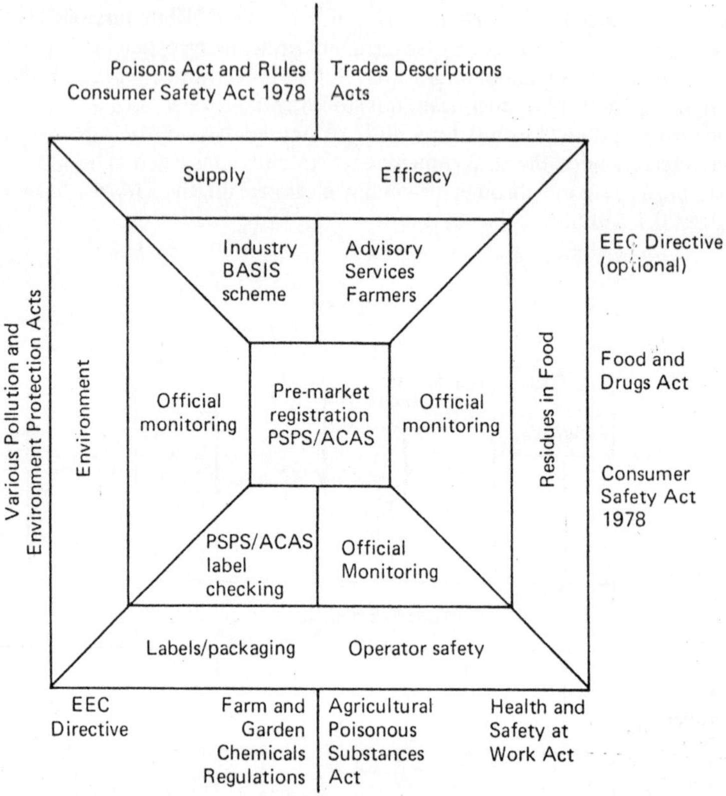

Figure 6.2 Showing the safeguards surrounding the use of agricultural chemicals in the UK (reproduced by permission of Ministry of Agriculture, Fisheries and Food, London). (EEC = European Economic Community, PSPS = Pesticide Safety Precautions Scheme, ACAS = Agricultural Chemicals Approval Scheme, BASIS = British Agrochemical Supply Industry Scheme Ltd)

arising from that use; and methods of analysis, medical data and uses in other countries. Initially the manufacturer or agent (notifier) is given trials clearance whereby small scale trials are undertaken to obtain primary data on efficacy and residue levels. This leads to limited clearance whereby a restriction is placed on the area to be treated or the amount of chemical that may be used by professional users such as the manufacturer's scientific staff and collaborators from research organisations and advisory services. Provisional commercial clearance is given for a period up to two years during which long-term animal testing for carcinogenicity is undertaken.

Finally, if the material has satisfied all criteria, full commercial clearance is given for unrestricted commercial sales although the material may be reviewed at any time in the light of new data. The extent of testing and monitoring new materials and new uses and the legislation that surrounds the UK clearance system is shown in figure 6.2. The clearance system finally designates the label to appear on the container; this gives instructions for rates of use, times and methods of application. These are precautions that will as far as possible ensure that the product is used safely. Protection is of three types.

1. *To the user*, detailing the protective clothing required, any specific hazards to be guarded against and the hygiene measures needed.
2. *To consumers*, giving details of crops on which the material may be used and the minimum time between last application and harvest.
3. *To livestock, wildlife and the environment*, especially hazards to birds, bees and fish, and giving advice on the storage of the product and on safe disposal of empty containers.

Closely linked with chemical clearance in the UK is an approval scheme to ensure that only effective chemicals are used in crop production. The *Agricultural Chemicals Approvals Scheme (ACAS)* is a voluntary scheme under which proprietary formulations of chemicals that have received PSPS clearance can be officially approved. It enables users to select and advisers to recommend efficient and appropriate chemicals and to discourage the use of unsatisfactory products. A *Certificate of Approval*, which is subject to annual renewal, is granted to each product and labels and literature may then display a large black 'A' surmounted by a crown. A list of approved products is published annually with supplements added during the year. This lists the proprietary names of approved products under the headings of the active ingredients they contain.

The use of products in commerce is monitored through a variety of agencies and the results, collated and published by survey groups, give estimates of the use of different chemicals on a range of crops, in food storage and food processing. Recently distributors have established the *British Agrochemical Supply Industry Scheme Limited (BASIS)* with the following objectives.

1. To maintain high standards of safety in the distribution chain.
2. To ensure that staff involved in distribution are knowledgeable particularly on matters concerning safety and efficacy.
3. To encourage safe use of agrochemicals within the consuming industries.

REFERENCES

ANON (1982a). *Approved Products for Farmers and Growers*, Ministry of Agriculture, Fisheries and Food, London

ANON (1982b). *The Scottish Agricultural Colleges Farm Management Handbook*, School of Agriculture, Aberdeen

BARNES, R. C., KLOTZ, L. J., de WOLFE, T. A., SMALL, R. H. and TURNER, G. U. (1966). Nematocidal and fungicidal properties of some soil fumigants. *Phytopathology*, 56, 691–8

BENT, K. J., COLE, A. M., TURNER, J. A. W. and WOLNER, M. (1971). Resistance of cucumber powdery mildew to dimethirimol. *Proceedings 6th British Insecticide Fungicide Conference*, 1, 274–82

BEN-YEPHET, Y., DINOOR, A. and HENIS, Y. (1975). The physiological basis of carboxin sensitivity and tolerance in *Ustilago hordei*. *Phytopathology*, 65, 936–42

BIGGS, G. E. and ROBERTSON, R. N. (1957). Apparent free space. *Annual Review Plant Physiology*, 8, 11–30

BROWN, M. E. and MELVILLE, S. C. (1970). Control of clubroot (*Plasmodiophora brassicae*) by soil chemicals. *Plant Pathology Laboratory, Harpenden, Monthly Summary* Nos 11 and 12, 55–6

BRYDE, R. J. W. (1969). Systemic fungicides including recent developments in the agricultural field. *Proceedings 5th British Insecticide Fungicide Conference*, 3, 675–9

BUCZACKI, S. T. and WHITE, J. G. (1979). The value of soil sterilants for the control of clubroot on a field scale. *Plant Pathology*, 28, 36–9

BURCHILL, R. T. and HUTTON, K. E. (1965). The suppression of ascospore production to facilitate the control of apple scab (*Venturia inaequalis* (Cke.) Wint.). *Annals Applied Biology*, 56, 285–92

CLARK, J., CLIFFORD, D. R., DEAS, A. H. B., GENDLE, P. and WATKINS, D. A. M. (1978). Photolysis metabolism and other factors influencing the performance of triadimefon as a powdery mildew fungicide. *Pesticide Science*, 9, 497–506

DAVIDSE, L. C. (1973). Antimitotic activity of methyl benzimidazol-2-yl carbamate (MBC) in *Aspergillus nidulans*. *Pesticide Biochemistry Physiology*, 3, 317–25

DAVIDSE, L. C. (1975a). Mode of action of methyl benzimidazol-2-yl carbamate (MBC) and the mechanism of resistance against this fungicide in *Aspergillus nidulans*. *Proceedings Symposium Systemic Fungicides*, May, Reinhardsbrunn, DDR, 137–43

DAVIDSE, L. C. (1975b). Antimitotic activity of methyl benzimidazol-2-yl carbamate in fungi and its binding to cellular protein. *Proceedings Symposium Microtubules and Microtubular Inhibitors*, Sept., Beerse, Belgium, 317–25

DEKKER, J. (1963). Antibiotics in the control of plant diseases. *Annual Review Microbiology*, 17, 243–62

DEKKER, J. (1967). Conversion of 6-azauracil in sensitive and resistant strains of *Cladosporium cucumerinum*. 333–9, in *Wirkungsmechanismen von Fungiziden und Antibiotika* (ed. M. Girbardt), Akademie Verlag, Berlin

DEKKER, J. (1971). Selective action of fungicides and development of resistance in fungi to fungicides. *Proceedings 6th British Insecticide Fungicide Conference*, 3, 715–23

DEKKER, J. (1977). Resistance. 176–197, in *Systemic Fungicides* (ed. R. W. Marsh), Longman, London

DEKKER, J. and OORT, A. J. P. (1964). Mode of action of 6-azauracil against powdery mildew. *Phytopathology*, 54, 815–8

DELP, C. J. and KLOPPING, H. L. (1968). Performance attributes of a new fungicide and mite ovicide candidate. *Plant Disease Reporter*, 52, 95–9
DIMOND, A. E. (1962). Surface factors affecting the penetration of compounds into plants. 368–82, in *Modern Methods of Plant Analysis*, Vol. 5 (eds H. E. Linskens and M. V. Tracey), Springer-Verlag, Berlin
EBBEN, M. H. and SPENCER, D. M. (1973). The integrated use of two systemic fungicides for control of cucumber powdery mildew. *Proceedings 7th British Insecticide Fungicide Conference*, 1, 211–16
EDINGTON, L. V. (1981). Structural requirements of systemic fungicides. *Annual Review Phytopathology*, 19, 107–24
EDINGTON, L. V., de WILDT, P. P. Q., JACQUES, K. and PSUTKA, J. (1978). The study of transcuticular movement of fungicides. *New York State Agricultural Experiment Station Special Report*, 28, 32–6
EDINGTON, L. V., MARTIN, R. A., BRUIN, G. C. and PARSONS, I. M. (1980). Systemic fungicides: a perspective after 10 years. *Plant Disease*, 64, 19–23
ELIAS, R. S., SHEPHARD, M. C., SNELL, B. K. and STUBBS, J. (1968). 5-n-Butyl-2-dimethylamino-4-hydroxy 6-methyl pyrimidine: a systemic fungicide. *Nature, London*, 219, 1160
EVANS, E. (1968). *Plant Diseases and their Chemical Control*, Blackwell, Oxford
EVANS, E. (1971). Systemic fungicides in practice. *Pesticide Science*, 2, 192–6
EVANS, E. (1977). Efficient use of systemic fungicides. 198–212, in *Systemic Fungicides* (ed. R. W. Marsh), Longman, London
FINNEY, J. R., FARRELL, G. M. and BENT, K. J. (1975). Bupirimate—a new fungicide for the control of powdery mildews on apples and other crops. *Proceedings 8th British Insecticide Fungicide Conference*, 2, 667–73
FROHBERGER, P. E. (1973). Bekämpfung von Pilzkrankheiten im Getreidebau durch Saatgutbehandlung. *Mitteilungen Biologischen Bundesanstalt Land-Forstwirtschaft, Berlin-Dahlem*, 151, 61–73
GARRETT, S. D. (1956). *Biology of Root Infecting Fungi*, Cambridge University Press, London
GASTONYI, M. and JOSEPOVITS, G. (1979). The activation of triadimefon and its role in the selectivity of fungicide action. *Pesticide Science*, 10, 57–65
GEORGOPOULOS, S. G. (1962). Tolerance to chlorinated nitrobenzenes in *Hypomyces solani* f *cucurbitae* and its mode of inheritance. *Phytopathology*, 53, 1086–93
GILPATRICK, J. D. and BLOWERS, D. R. (1974). Ascospore tolerance to dodine in relation to orchard control of apple scab. *Phytopathology*, 64, 649–52
GREENAWAY, W. (1971). Relationship between mercury resistance and pigment production in *Pyrenophora avenae*. *Transactions British Mycological Society*, 56, 37–44
GREWE, F. and BÜCHEL, K. H. (1973). Ein neues Mehltaufungizid aus der Klasse der Trityltriazole. *Mitteilungen Biologischen Bundesanstalt Land-Forstwirtschaft, Berlin-Dahlem*, 151, 208–9
HESS, C. and LÖCHER, F. (1975). Experiences with vinclozolin in the control of *Botrytis cinerea* in strawberries. *Proceedings 8th British Insecticide Fungicide Conference*, 2, 693–6
HUANG, K. T., MISATO, T. and ASUYAMA, H. (1964). Selective toxicity of Blasticidin-S to *Piricularia oryzae* and *Pellicularia saskii*. *Journal Antibiotics*, Series A, 17, 71–4
JACKS, H. (1953). Seed disinfection. V. A seed dressing apparatus for the smaller seedsman. *New Zealand Journal Science Technology*, B34, 249–51
KAARS SIJPESTEIJN, A. and VAN DER KERK, G. J. M. (1970). Biochemical modes of action of fungicides. *Proceedings 5th British Insecticide Fungicide Conference*, 3, 724–33

KAPPAS, A. and GEORGOPOULOS, S. G. (1970). Genetic analysis of dodine resistance in *Nectria haematocca* (syn. *Hypomyces solani*). *Genetics*, **66**, 617-22

KELLEY, W. D. (1980). Evaluation of systemic fungicides for control of *Cronartium quercuum* f. sp. *fusiforme* on Loblolly pine seedlings. *Plant Disease*, **64**, 773-5

KIRBY, A. H. M. (1972). Progress towards systemic fungicides. *Pesticides Abstracts News Summary, Section B, Fungicides Herbicides*, 18, 1-33

KRADEL, J. and POMMER, E. H. (1967). Some remarks and results on the control of powdery mildew in cereals. *Proceedings 4th British Insecticide Fungicide Conference*, 1, 170-5

KRADEL, J., POMMER, E. H. and EFFLAND, H. (1969). Response of barley varieties to the control of powdery mildew with cyclomorph and tridemorph. *Proceedings 5th British Insecticide Fungicide Conference*, 1, 16-19

KRAMER, P. J. (1957). Outer space in plants, some possible implications of the concept. *Science New York*, 125, 633-5

LACROIX, L., BIC, G., BURGAUD, L., GUILLOT, M., LEBLANC, R. and SAULI, M. (1974). Étude des propriétés antifongiques d'une nouvelle famille de derivés de l'hydantoine et en particulier du 26109 R.p. *Phytiatrie-Phytopharmacie*, 23, 165-74

LAMBERT, D. H. and WUEST, P. J. (1975). Increased sensitivity to zineb for *Verticillium malthousei* strains tolerant to benomyl. *Phytopathology*, 65, 637-8

LUKENS, R. J. (1971). *Chemistry of Fungicidal Action*, Springer-Verlag, Berlin

MARTIN, H. (1964). *The Scientific Principles of Crop Protection*, St. Martin's Press, New York

MATHRE, D. E. (1970). Mode of action of oxathiin systemic fungicides. I. Effect of carboxin and oxycarboxin on the general metabolism of several Basidiomycetes. *Phytopathology*, 60, 671-6

MATTHEWS, G. A. (1979). *Pesticide Application Methods*, Longman, London

McCALLAN, S. E. A. (1967). History of Fungicides. 1-38, in *Fungicides—An Advanced Treatise*, vol. 1 (ed. D. C. Torgeson), Academic Press, New York

MILLARDET, A. and GAYON, U. (1885). Traitment du mildiou par le mélange de sulphate de cuivre et de chaux. *Journal Agriculture Pratique* (Paris), 49, 707-10 (English translation by F. J. Schneiderhan (1933). *Phytopathological Classics*, 3, 12-17)

NAKANISHI, T. and OKU, H. (1969). Metabolism and accumulation of pentachloronitrobenzene by phytopathogenic fungi in relation to selective toxicity, *Phytopathology*, 59, 1761-2

PAGE, G. E. (1963). Soil fumigation equipment. *Oregon State University Extension Bulletin*, 813, 1-14

PRICE, C. E. (1977). Movement of xenobiotics in plants. 42-66, in *Herbicides and Fungicides—Factors affecting their Activity* (ed. N. R. McFarlane), Chemical Society Special Publication No. 29, London

RICH, S. (1956). Foliage fungicides plus glycerin for the chemotherapy of cucumber scab. *Plant Disease Reporter*, 40, 620-1

SBRAGIA, R. J. (1975). Chemical control of plant diseases: an exciting future. *Annual Review Phytopathology*, 13, 257-69

SCHICKE, P. and VEEN, K. H. (1969). A new systemic, CELA W 524 (N, N'-bis-(1 formamide-2,2,2-trichloro ethyl)-piperazine) with action against powdery mildew, rust and apple scab. *Proceedings 5th British Insecticide Fungicide Conference*, 2, 569-75

SHARVELLE, E. G. (1961). *The Nature and Uses of Modern Fungicides*, Burgess Publishing Co., Minneapolis

SOMERS, E. (1967). The sites of reaction of fungicides in spores. *Proceedings 18th Symposium Colston Research Society*, University of Bristol, 1966, 299-308

SZKOLNIK, M. and GILPATRICK, J. D. (1973). Tolerance of *Venturia inaequalis* to dodine in relation to the history of dodine usage in apple orchards. *Plant Disease Reporter*, 57, 817-21

THORN, G. D. and LUDWIG, R. A. (1962). *The Dithiocarbamates and Related Compounds*, Elsevier, Amsterdam

URECH, P. A., SCHWINN, F. and STAUB, T. (1977). CGA 48988, a novel fungicide for the control of late blight, downy mildews and related soil-borne diseases. *Proceedings 9th British Insecticide Fungicide Conference*, 2, 623-31

VAN DER KERK, G. J. M. (1971). Systemic fungicides, new solutions and new problems. *Proceedings 6th British Insecticide Fungicide Conference*, 3, 791-802

VON SCHMELING, B. and KULKA, M. (1966). Systemic fungicidal activity of 1,4-oxathiin derivatives. *Science*, New York, 152, 659-60

WAIN, R. L. and CARTER, G. A. (1977). Nomenclature and Definitions, 1-5, and Historical Aspects, 6-31, in *Systemic Fungicides* (ed. R. W. Marsh), Longman, London

WATKINS, T. C. and NORTON, L. B. (1955). *Handbook of Insecticide Dust Diluents and Carriers* (revised by D. E. Weidhaas and J. L. Brann), Dorland Books, Caldwell, New Jersey

WILHELM, G. and NELSON, P. E. (1970). A concept of rootlet health of strawberries in pathogen free soil achieved by fumigation. 208-15, in *Root Diseases and Soil Borne Pathogens* (eds T. A. Toussoun, R. V. Bega and P. E. Nelson), California University Press, Berkeley

WOLFE, M. S. and DINOOR, A. (1973). The problem of fungicide tolerance in the field. *Proceedings 7th British Insecticide Fungicide Conference*, 1, 11-19.

FURTHER READING

ALLEN, M. W. (1960). Nematicides. 604-38, in *Plant Pathology—An Advanced Treatise*, Vol. III, *The Pathogen* (eds J. G. Horsfall and A. E. Dimond), Academic Press, New York

ANON (1982). *Pesticide Safety Precautions Scheme*, Ministry of Agriculture, Fisheries and Food, London

ANON (1982). *Pesticides Safety Precautions Scheme, Products Cleared for Agricultural, Food Storage, Public Hygiene, Domestic and Related Uses in the United Kingdom*, Ministry of Agriculture, Fisheries and Food, London

BRANDES, G. A. (1971). Advances in fungicide utilization. *Annual Review Phytopathology*, 9, 363-86

BURCHFIELD, H. P. (1960). Performance of fungicides on plants and in soil—physical, chemical and biological considerations. 477-520, in *Plant Pathology—An Advanced Treatise*, Vol. III, *The Pathogen* (eds J. G. Horsfall and A. E. Dimond), Academic Press, New York

COURSHEE, R. J. (1967) Application and use of foliar fungicides. 240-87, in *Fungicides—An Advanced Treatise*, Vol. I, *Agricultural and Industrial Applications, Environmental Interactions* (ed. D. C. Torgeson), Academic Press, New York.

CROSS, J. E. (1971). Prospects for the use of bactericides for the control of bacterial diseases. *Proceedings 6th British Insecticide Fungicide Conference*, 3, 694-705

DEKKER, J. (1976). Acquired resistance to fungicides. *Annual Review Phytopathology*, 14, 405-28

DEUTSCH, A. E. and POOLE, A. P. (eds) (1972). *Manual of Pesticide Application Equipment*. International Plant Protection Center Oregon State University Corvallis

DIMOND, A. E. (1962). Objectives in plant chemotherapy. *Phytopathology*, 52, 1115–18

ERWIN, D. C. (1973). Systemic fungicides: disease control, translocation and mode of action. *Annual Review Phytopathology*, 11, 389–422

EVANS, E. (1971). Systemic fungicides in practice. *Pesticide Science*, 2, 192–6

EVANS, E. (1971). Problems and progress in the use of systemic fungicides. *Proceedings 6th British Insecticide Fungicide Conference*, 3, 758–64

FOX, R. T. V. (1971). Development of *in vitro* and *in vivo* screening procedures for bactericides. *Proceedings 6th British Insecticide Fungicide Conference*, 1, 187–92

FUCHS, A. (ed.) (1977). Internal therapy of plants. *Netherlands Journal Plant Pathology*, 83 (Supplement)

GARRETT, S. D. (1960). Inoculum potential. 23–57, in *Plant Pathology, An Advanced Treatise*, Vol. III, *The Diseased Population, Epidemics and Control* (eds J. G. Horsfall and A. E. Dimond), Academic Press, New York

GOODMAN, R. N. (1962) The impact of antibiotics upon plant disease control. *Advances in Pest Control Research*, 5, 1–46

GORING, C. A. I. (1962). Theory and principles of soil fumigation. *Advances in Pest Control Research*, 5, 47–84

KIRBY, A. H. M. (1972). Progress towards systemic fungicides. *Pesticides Abstracts News Summary Section B Fungicides Herbicides*, 18, 1–33

KREUTZER, W. A. (1960). Soil Treatment. 431–76, in *Plant Pathology–An Advanced Treatise*, Vol. II, *The Pathogen* (eds J. G. Horsfall and A. E. Dimond), Academic Press, New York

LEAR, B. (1951). Use of methyl bromide and other volatile chemicals for soil fumigation. *Cornell University Agricultural Experiment Station*, New York, Memoir 303

LUDWIG, R. A. and THORN, G. D. (1960). Chemistry and mode of action of dithiocarbamate fungicides. *Advances in Pest Control Research*, 3, 219–52

LUKENS, R. J. (1971). *Chemistry of Fungicidal Action*, Springer-Verlag, Berlin

MARSH, R. W. (ed.) (1977). *Systemic Fungicides*, Longman, London

MARTIN, H. and WORTHING, C. R. (eds) (1976). *Insecticide and Fungicide Handbook*, 5th edn, Blackwell, London

McFARLANE, N. R. (ed.) (1977). *Herbicides and Fungicides–Factors affecting their Activity*, Special Publication No. 29, The Chemical Society, London

McKENRY, M. V. (1978). Selection of preplant fumigation. *California Agriculture*, 32 (1), 15–16

McKENRY, M. V. and THOMASON, I. J. (1974). 1,3-Dichloropropene and 1,2-dibromoethane compounds: I. Movement and fate as affected by various conditions in several sites. II. Organism–dosage–response studies in the laboratory with several nematode species. *Hilgardia*, 42, 393–438

MULDER, D. (ed.) (1979). *Soil Disinfestation*, Elsevier, Amsterdam

MUNNECKE, D. E. (1972). Factors affecting the efficacy of fungicides in soil. *Annual Review Phytopathology*, 10, 375–98

MUNNECKE, D. E. and VAN GUNDY, S. D. (1979). Movement of fumigants in soil, dosage responses and differential effects. *Annual Review Phytopathology*, 17, 405–29

MUNNECKE, D. E., LUDWIG, R. A. and SAMPSON, R. E. (1959). The fungicidal activity of methyl bromide. *Canadian Journal Botany*, 37, 51–8

PURDY, L. H. (1967). Application and use of soil and seed treatment fungicides. 195–239, in *Fungicides–An Advanced Treatise*, Vol. I, *Agricultural and Industrial Applications, Environmental Interactions* (ed. D. C. Torgeson), Academic Press, New York

RICH, S. (1960). Fungicidal Chemistry. 553–603, in *Plant Pathology–An Advanced*

Treatise, Vol. II, *The Pathogen* (eds J. G. Horsfall and A. E. Dimond), Academic Press, New York

RODRIGUEZ-KABANA, R., BACKMAN, P. A. and CURL, E. A. (1977). Control of seed and soil-borne plant diseases. 117–61, in *Antifungal Compounds.* Vol. 1 (eds M. R. Siegel and H. D. Sisler), Dekker, New York

SHEPHERD, M. C. (1972). Factors limiting the systemic activity of fungicides. 383–88, in *Herbicides, Fungicides, Formulation Chemistry* (ed. A. S. Tahori), Gordon and Breach, New York

SISLER, H. D. and COX, C. E. (1960). Physiology of Fungitoxicity. 507–552, in *Plant Pathology—An Advanced Treatise*, Vol. II. *The Pathogen* (eds J. G. Horsfall and A. E. Dimond), Academic Press, New York

SOMERS, E. (1967). Formulation. 153–94, in *Fungicides—An Advanced Treatise*, Vol. I, *Agricultural and Industrial Applications, Environmental Interactions*, (ed. D. C. Torgeson), Academic Press, New York

TORGESON, D. C. (ed.) (1969). *Fungicides—An Advanced Treatise*, Vol. II, *Chemistry and Physiology*, Academic Press, New York

VAARTAJA, O. (1964). Chemical treatment of seedbeds to control nursery diseases. *Botanical Review*, 30, 1–91

VAN DER KERK, G. J. M. (1977). Evolution of the chemical control of plant diseases. *Netherlands Journal Plant Pathology*, **83** (supplement 1), 3–13

VAN GUNDY, S. D. and McKENRY, M. V. (1977). Action of nematicides. 263–83, in *Plant Disease—An Advanced Treatise*, Vol. I. *How Disease is Managed* (eds J. G. Horsfall and E. B. Cowling), Academic Press, New York

VAN TIEL, N. (1975). Principles and problems of pesticide legislation, *Pesticide Science*, 6, 189–97

VERLOOP, A. (1972). Factors influencing the availability of systemic fungicides in plants. 389–96, in *Herbicides, Fungicides, Formulation Chemistry* (ed. A. S. Tahori), Gordon and Breach, New York

WAIN, R. L. and CARTER, G. A. (1967). Uptake, translocation and transformation by higher plants. 561–612, in *Fungicides—an Advanced Treatise*, Vol. I, *Agricultural and Industrial Applications, Environmental Interactions* (ed. D. C. Torgeson), Academic Press, New York

WELLMAN, R. H. (1977). Problems in development, registration and use of fungicides. *Annual Review Phytopathology*, 15, 153–63

WILHELM, S., STORKAN, R. C. and WILHELM, J. M. (1974). Preplant soil fumigation with methyl bromide–chloropicrin mixtures for control of soil-borne diseases of strawberries—a summary of fifteen years of development. *Agriculture Environment*, 1, 227–36

WOODCOCK, D. (1977). Chemical structure and fungicide selectivity. *Netherlands Journal Plant Pathology*, **83** (Supplement 1), 27–37

WORTHING, C. R. (ed), (1979). *The Pesticide Manual—A World Compendium*, British Crop Protection Council, Croydon

ZENTMEYER, G. A., WALLACE, P. P. and HORSFALL, J. G. (1944). Distance as a dosage factor in the spread of Dutch elm disease. *Phytopathology*, 34, 1025–33

7 CONTROL: SPRAY APPLICATION

Spraying is the predominant method by which chemicals are used to control plant pathogens (section 6.4). This chapter describes, in non-engineering terms, the major components of a sprayer and some factors that affect their effectiveness. Basically a sprayer is a mechanical device whereby the manufacturer's product is diluted by mixing with water or other fluid in a reservoir vessel and then forced by hydraulic or air pressure through one or more nozzles. These break up the liquid into droplets which are impelled to a target crop. Choice of sprayer depends on the total volume of spray to be applied, the area of crop to be covered, the habit of the crop—some sprayers are suited to low-growing vegetable crops rather than to trees or bushes—and crop location, that is, whether the crop is grown under protection or in the open field.

The main types of sprayer, based on their means of propulsion, are as follows.

1. *Hand-held*, which apply small volumes of spray to limited areas and are mostly suited to protected crops and for domestic purposes.
2. *Knapsack-mounted or pedestrian-propelled machines*, which may be manually operated, driven by an electric motor or by a small combustion engine; these are suitable for the application of small volumes of liquid to restricted areas either in the field or under protection.
3. *Tractor-mounted or trailed ground-crop sprayers*, which are fitted with an engine-powered pump for field use and apply sprays at low, medium or high volume (section 7.2.4) according to jet size and pump output.
4. *Orchard sprayers*, which are mainly of the air-carrier type applying medium-volume and low-volume sprays.
5. *Aircraft sprayers*, which are used for extensive areas of monocrops and are particularly useful in difficult terrain or where tractor-mounted sprayers would either cause crop damage, and consequent yield losses, or are incapable of providing effective spray distribution.

7.1 MAIN COMPONENTS OF CONVENTIONAL SPRAYERS

These are outlined for hydraulic sprayers in figure 7.1.

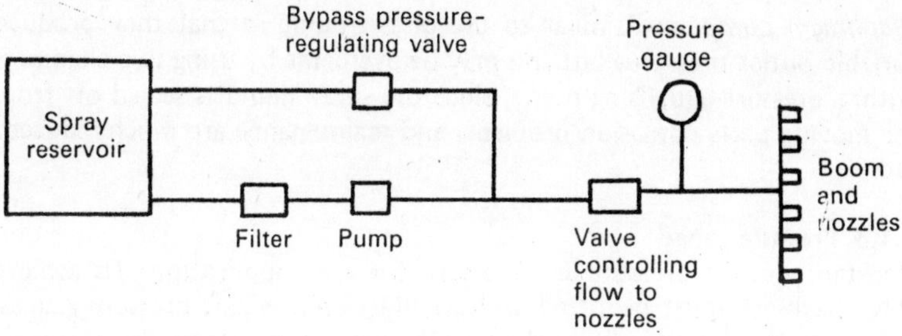

Figure 7.1 Basic components of a hydraulic sprayer

7.1.1 Tank

Reservoir tanks are generally constructed of galvanised steel, polyethylene or glass reinforced plastic (glass fibre). In order to reduce weight and prevent corrosion the last two are now most popular. Sprayers capable of applying greater than 25 l require some provision for agitation of the spray to ensure constant mixing. This is achieved either mechanically by a propeller or more rarely with stirrer blades, or hydraulically whereby the pump draws excess spray from the tank, provides sufficient to keep the nozzles supplied and returns the surplus via a bypass valve back to the tank causing continuous agitation. A filter is fitted between the tank and pump to prevent solids, which may come from the farm water supply, from causing damage to the pump or blocking the nozzles. When wettable powder formulations are used this filter must be coarse enough to permit passage of the suspended chemical particles.

7.1.2 Pumps

Roller-vane pumps are suited to small and medium-sized machines and are probably the most widely used type. They are more expensive than a comparable gear pump but less susceptible to wear caused by spray suspensions.

Gear pumps are used on small sprayers but suffer from a high rate of wear when pumping suspensions.

Centrifugal pumps are suited to small sprayers but are relatively expensive, owing to pressure limitations of a single-stage pump, and hence they are usually only fitted to large high-volume machines; they require little maintenance.

Piston pumps are of two types. They can be either single-cylinder ones that produce very wide variation in output pressure and are therefore undesirable for any machine other than hand-held sprayers used on small areas, or multi-cylinder ones that are useful in situations where sprays need to be applied at high pressures.

Diaphragm pumps are similar to the piston pump in that they produce variable outlet pressures but this may be overcome by using two chambers with a pressure-equalising dome, since the spray liquid is sealed off from the moving parts corrosion problems and maintenance are much reduced. reduced.

7.1.3 Pressure gauge

Constant output pressure is necessary for even application. To achieve this, machines must be fitted with accurate and reliable pressure gauges. These are also essential for calibrating the machine to ensure that a correct rate of application is achieved. Considerable wear and tear occurs to gauges in the course of a season and they need to be replaced regularly.

7.1.4 Nozzles

Anvil (figure 7.2) The spray is directed through the nozzle body and impacts against a curved anvil which causes it to break up into droplets. These nozzles produce large variable-sized droplets in an uneven spray pattern. The orifice size of a particular nozzle is fixed but several sizes are available.

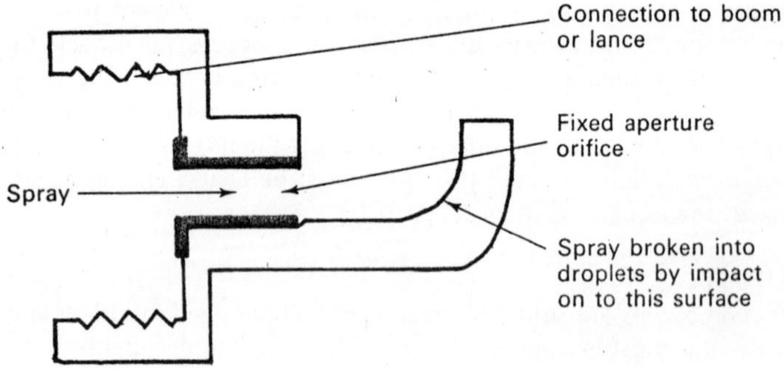

Figure 7.2 Diagram of a cross-section of an anvil nozzle

Fan (figure 7.3) Various sizes of fan nozzle are available and are designated by letter or number codes. The nozzle body fixes to the boom or lance and within this is a filter and detachable metal or ceramic jet tip, both being secured by a locking nut. This type of nozzle produces a flat fan of spray (figure 7.7) and when they are set up along the boom all fans should be at a slight angle to avoid interference between them. This form of nozzle produces a variable droplet size with a proportion of very small spray droplets which are likely to drift. This characteristic can be minimised by using larger-sized nozzles. Spray pattern is uneven across the fan; it is compensated for by overlapping nozzle output.

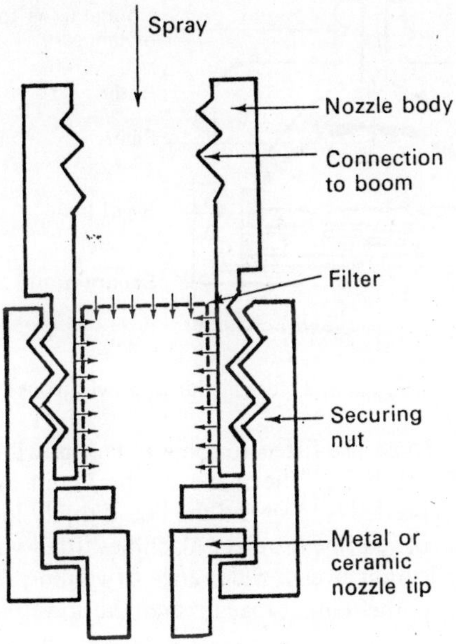

Figure 7.3 Diagram of a cross-section of a fan nozzle

Swirl (cone) (figure 7.4) These consist of a metallic or plastic body which is screwed into the boom or lance and within this are two discs separated by a washer, thereby forming a swirl chamber. The parts are held into the body by a securing nut. The inner disc has a series of slanting holes drilled through it so that a swirling motion is imparted to the spray as it passes through. The spray pattern produced from this type of nozzle is a hollow cone. The angle of the cone and fineness of spray depend on the swirling velocity, dimensions of the swirl chamber and output of the nozzle. The outer disc has a central hole the size of which, together with pump pressure, governs nozzle output. Nozzles can be designed for either most efficient break-up of spray (high swirl rate at a wide angle) or to give a coarser narrow-angled cone of spray which has a higher velocity and will travel further. Large droplets can have a greater momentum and be projected further than small ones. This is an important consideration with hydraulic sprayers where the spray must not be produced more finely than can be projected into a crop at the pressure available. Where hand-held sprayers and lances are used, a form of nozzle is available in which the cone angle can easily be adjusted during use. A wide-angle spray may then be used where conditions permit a fine spray, or the spray pattern may be narrowed and coarsened to provide greater penetration.

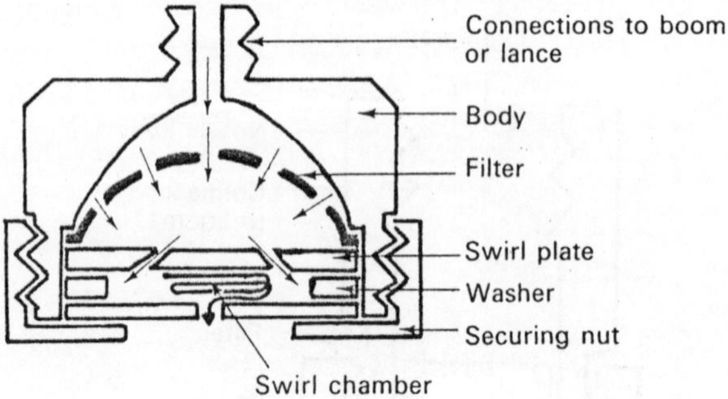

Figure 7.4 Diagram of a cross-section of a swirl (cone) nozzle

Shear (figure 7.5) These are fitted to some tractor-powered and powered knapsack air-carrier machines. The shape of the liquid outlet varies but essentially all that is needed is to spread the liquid thinly into an air stream so that it will shear off as fine droplets. Machines fitted with such nozzles are capable of application over a wide range of volumes, the upper limit being determined by the tank capacity and the lower limit by the air pressure available.

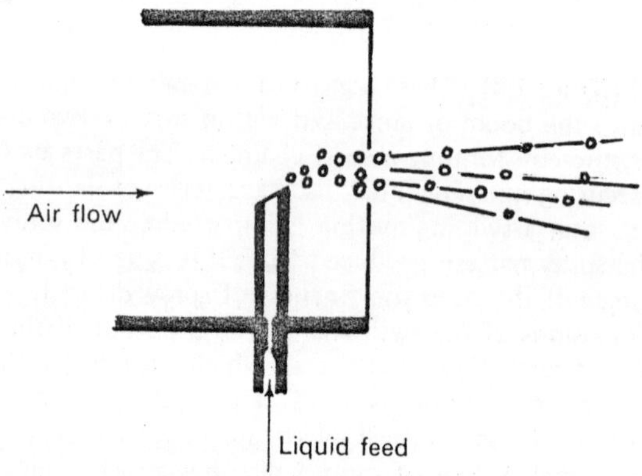

Figure 7.5 Diagram of a cross-section of a shear nozzle

The distribution of spray produced by a nozzle or group of nozzles may be measured with a *patternator* (figure 7.6). Several types are available; at the simplest they consist of a deeply corrugated sloping metal tray, with the nozzle(s) mounted about 45 cm above the tray. Liquid is sprayed

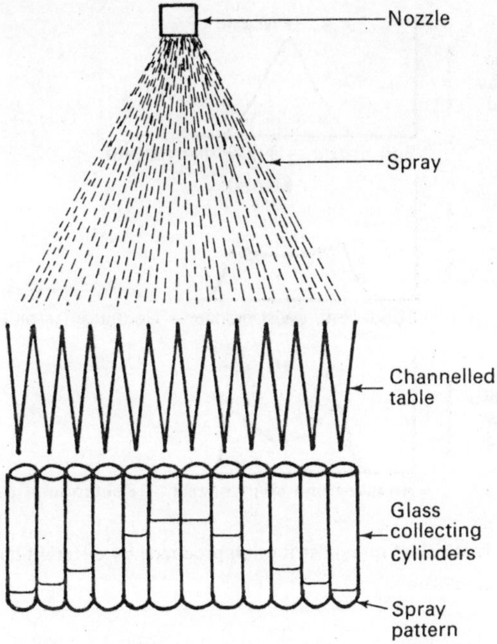

Figure 7.6 Diagram of apparatus used to determine the pattern of spray-distribution from a nozzle (patternator)

on to the tray, gathered in the corrugations and eventually collected in a series of graduated glass cylinders. The forms of liquid distribution produced by the most popular nozzles, fan and cone, are shown in figure 7.7. Fan nozzles produce a triangular spray pattern while cones produce a rectangular pattern. The characteristics of these nozzles are given in table 7.1.

7.1.5 Booms

Most booms are constructed of plastic tubing and vary in width up to 20 m. Rate of work increases and the amount of crop damage due to tractor-wheelings decreases with wider booms. Wide booms make it more difficult to match spray swaths accurately and application across individual swaths tends to be less uniform because of variations in boom height (whip and bounce) caused by tractor motion. The boom must be sufficiently rigid to minimise whip. Booms may now be carried on a gimbal-type suspension often with hydraulic damping to minimise whip and bounce. The tube diameter must be wide enough to prevent pressure variation between

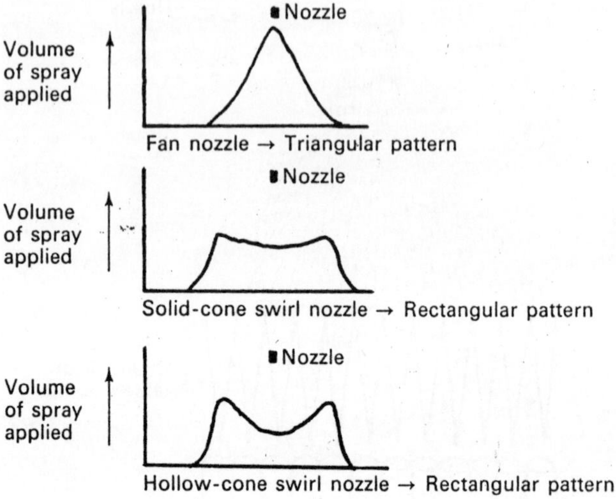

Figure 7.7 Patterns of spray-distribution produced by different nozzle types

Table 7.1

The main characteristics of hollow-cone and fan-jet nozzles (after Evans, 1968)

	Hollow-cone nozzles	*Fan-jet nozzles*
1.	Liquid rotates within the nozzle and the spray lacks forcefulness; they are best for spraying leaf surfaces but are liable to give rise to spray drift	Straight-through flow results in a high kinetic-energy spray with good penetration; it is suitable for spraying deep into crop canopies
2.	Hollow-cone patterns are difficult to overlap and thereby achieve uniform output along the spray bar but the operational height of the bar is not critical	Fan jets can be designed to produce a relatively uniform output along the spray bar, but bar height is of critical importance
3.	These nozzles produce accurate output rates and are relatively hard wearing	Output from fan jets is unpredictable and tends to increase rapidly with wear
4.	By means of a central hole in the swirl plate these nozzles can produce a coarse forceful spray which is useful for plantation crops	The main function is for spraying low-growing annual field crops

nozzles at the middle and end of the boom; alternatively the boom may be fed at several points. For spraying densely standing field crops such as Brussels sprout or bush crops such as blackcurrant, *dropleg* booms may be used. These are vertical attachments fitted to a conventional boom, with two or more nozzles per leg to project spray beneath and upwards into the canopy. The nozzles, usually of the hollow-cone type, are set parallel to their line of travel. For very low-growing crops such as strawberry, an inverted 'T'-shaped attachment may be fitted to allow the nozzles to spray upwards and place spray beneath the foliage. Alternatively a low-set conventional boom is used but this is fitted with a *ruffler* or *roller bar* to agitate the foliage and improve spray penetration, as demonstrated in table 7.2.

Table 7.2

Spray penetration of a strawberry canopy in the presence and absence of rufflers (after Cole, 1978)

Ruffler type	*Percentage of leaf area covered*	
	Upper-leaf surface	*Lower-leaf surface*
Without ruffler	77	34
Using solid bar	67	58
Using chain	95	71

Another method used to ensure more effective and economical use of chemicals is *band-spraying*. Chemicals are applied in distinct narrow bands to low-growing row crops such as strawberries and field vegetables with considerable saving in the quantity of spray required per unit area of land. Usually nozzles producing a rectangular spray pattern, such as the hollow cone, are fitted although fan jets may be used. In this case the fans are set at an angle of 45° to the line of travel rather than at 90°. The most satisfactory band width is about 180 mm.

All booms are equipped with a means of height adjustment and it is essential to set this correctly to achieve even coverage of a target crop. Figure 7.8 illustrates the correct adjustment of booms when fitted with nozzles producing triangular and rectangular spray patterns. To compensate for boom oscillation (whip) during motion, it may be necessary to raise the boom by an additional distance equal to half the vertical oscillation of the boom when travelling (figure 7.9). In tall crops a compromise on boom height is inevitable. It is easier, however, to obtain good coverage with a boom set too high rather than too low. It must always be borne in mind that raising boom height increases the risk of spray drift.

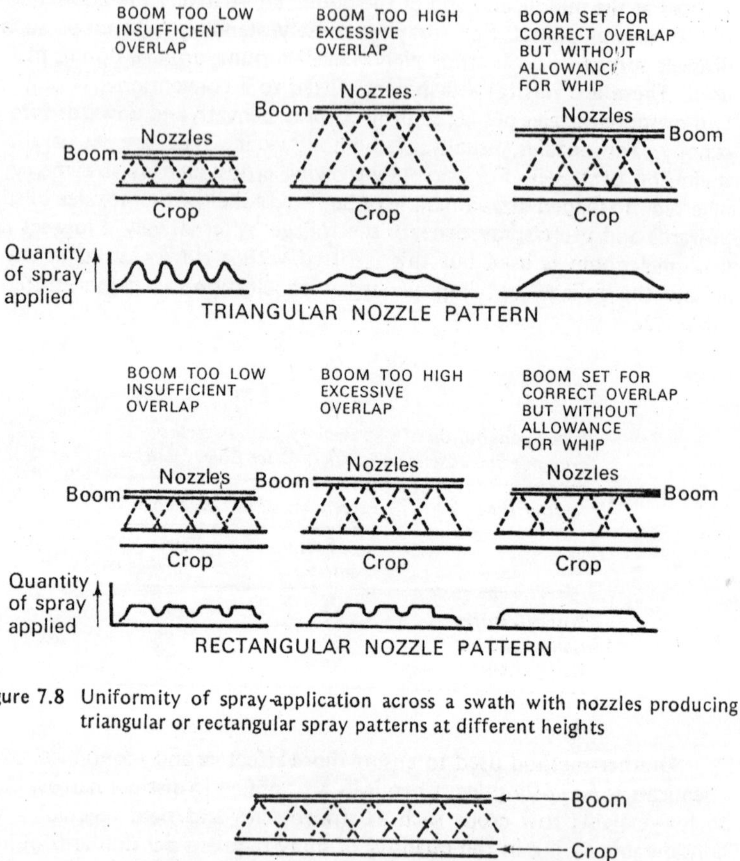

Figure 7.8 Uniformity of spray-application across a swath with nozzles producing triangular or rectangular spray patterns at different heights

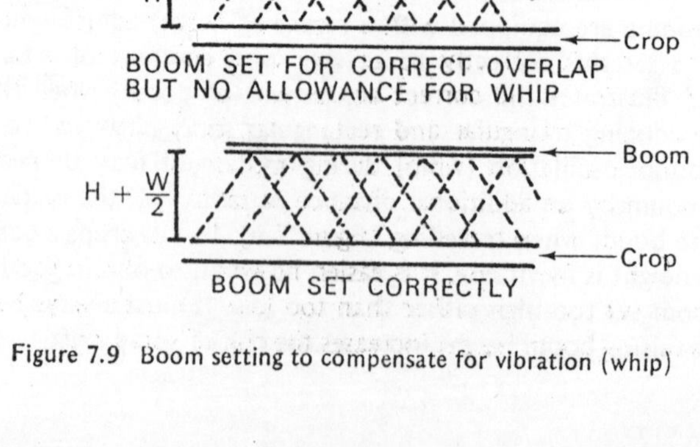

Figure 7.9 Boom setting to compensate for vibration (whip)

7.1.6 Fans

Sprayers for plantation tree crops utilise air-carrier fans to provide sufficient velocity to the spray to carry it into the trees. In most types of machine the spray is delivered by a pump to one or more nozzles and then carried by an air stream generated by one or more fans. The effective range of the air stream and in some cases the size of the spray droplets depends on air velocity and output. Both axial-flow and centrifugal fans are fitted to air-carrier sprayers. Axial-flow fans cannot be used with shear nozzles but centrifugal fans may be used in conjunction with all nozzle types.

7.2 IMPROVING THE EFFECTIVENESS OF SPRAYING

7.2.1 Objectives

The objective of applying crop-protection chemicals is to distribute as small as possible a quantity of active ingredient in order that it will have maximum effect in preventing the establishment or progress of an infection. With protectant and systemic materials this may mean spraying in advance of invasion to prevent infections which occur with predictable regularity. It may not be essential, for example, to apply chemical to the whole plant in order to prevent the initiation of invasion. Ascospores of *Venturia inaequalis* (apple scab—section 2.6.5) are deposited primarily on the sepals and pedicels of developing flowers (Keitt and Jones, 1926) and margins of unfurling leaves (Moore, M. H., 1964). The lesion loci of *Phytophthora infestans* (potato blight—section 2.6.2) occur in significantly greater proportions at the tips and margins of leaves (Lapwood, 1961). Eradicant materials need to be placed where pathogen growth is most active. Consequently, for control of downy mildew fungi it is necessary to spray the leaf undersurfaces rather than covering the entire plant surface.

A general trend with spray application is to reduce the volume of water in which the chemical product is applied. Considerable thought is being given to means by which smaller volumes of spray may be used with greater effectiveness in controlling pathogens. This trend results from factors such as the high cost of transporting large volumes of spray through crops, the damage that can result to crops and soil structure from spraying operations using large volumes of water and the need to reduce the quantities of expensive active chemicals required for financial and environmental reasons. Reducing volume makes a spray programme more flexible in relation to weather conditions and cheaper per unit area treated. Changes in sprayer design are beginning to come forward but the basic mechanical principles of hydraulic sprayers have altered little over the last 100 years. The most significant advance is the progressive use over the last 30 years of air-carrier machines for plantation tree crops. Knapsack sprayers were used to apply fungicides to French vineyards in the nineteenth century (Lodeman, 1910) and their design was similar to those used today except that the components are now made of various plastics rather than metal (Matthews, 1981). Considerable problems still exist with sprayers failing to provide sufficiently even distribution of spray and penetration into the crop canopy. Most sprayers involve considerable waste through spray failing to reach the target plant and more particularly the target organisms which need to be controlled, leading to unnecessary environmental contamination. Neither high-volume nor low-volume spraying is very efficient in distributing chemicals uniformly or confining them to the target as demonstrated by Morgan (1981) and figure 7.10. Air-carrier low-volume spraying has been shown with bush apples, for instance, to distribute more fungicide to those leaf surfaces at the base of a tree, while insufficient reaches the inner and higher parts.

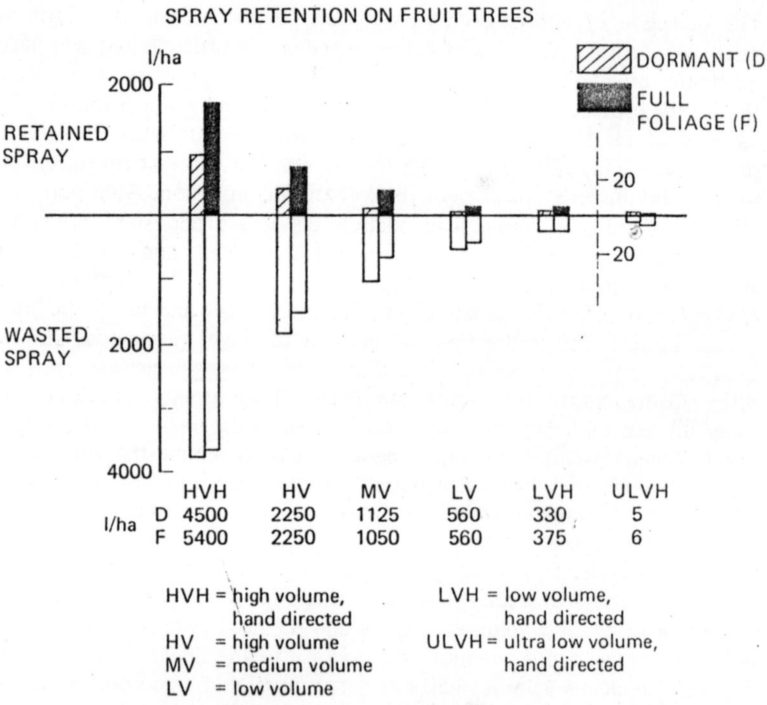

Figure 7.10 Spray retention on apple trees, illustrating the relative amounts of spray retained and wasted in the dormant and full-foliage seasons when applied by various methods. (Reproduced by permission of Dr. N. G. Morgan, Long Ashton Research Station and the Agricultural Research Council)

Alternatively, in attempts to cover the top of a tree adequately, the outer and lower parts receive excessive chemical.

7.2.2 Machine calibration

The most direct method of reducing wasteful spraying is by regular calibration of the machine. At its most basic level this consists of timing the sprayer over a measured distance, such as 100 m, and spraying out plain water at a speed of travel suitable for the job. The tank is then refilled and, with the sprayer stationary, water is sprayed for precisely the length of time required to travel 100 m. The tank is refilled and the volume of water required is measured. This volume and the swath width are checked against a calibration chart supplied with each spray. If output is in error, small changes may be made by adjusting spray pressure; larger modifications are made by selecting alternative nozzles. Details of tractor settings, sprayer volumes, pressures and timings are logged for future reference. More detailed methods for calibrating sprayers are given in Martin and Worthing (1976) and Anon (1980).

7.2.3 Retention and drift

An ideally effective application of chemical is one in which only the minimum amount is used, all being deposited and retained on the target with none escaping to non-target areas as contaminant waste. Spray that escapes from the target is termed *drift* and may be of two types.

1. *Exodrift* This is spray that is blown out of the target area and usually consists of small droplet sizes; these are easily picked up by air turbulence.
2. *Endodrift* This is spray that remains within the target area, running off into the surrounding soil, and is usually composed of large droplet sizes.

Spraying to run-off ensures complete wetting of the foliage but may be counter-productive since once run-off starts then retention of chemical on the foliage is less than if spraying stops just before run-off begins. Endodrift may also occur as a result of chemical being washed from foliage by rain or irrigation.

Droplets of spray are collected on foliage by the processes of sedimentation and impaction. Droplets larger than 150 μm diameter may possess sufficient kinetic energy to bounce away from the target following impact (Brunskill, 1956). The physical forces that bind chemicals to plant leaves include London van der Waal forces, electrostatic attraction and capillary action (Burchfield, 1959). Spray droplet retention is influenced also by properties of the target such as the following.

1. *Leaf shape and angle* Plants having large flat leaves held at right angles to a stem will retain spray droplets more easily than those with lanceolate leaves held at an acute angle.
2. *Leaf age* Cotyledons, newly opened and very old leaves are more easily wetted and retain spray to a greater extent than semi-mature and mature leaves. This may be related to changes in leaf-wax composition and structure and hairiness at different stages in plant growth. Properties of leaf wax affect the *angle of contact* of a spray droplet with the leaf and on this basis leaf surfaces may be divided into two groups. Where the angle of contact of the droplet is below 90°, superficial wax does not affect retention; where this angle exceeds 90°, wax significantly affects wetting. Contact angles of 90°–110° occur on leaves with a smooth layer of superficial wax.

7.2.4 Volume and droplet size

The terms and approximate volume rates currently used to describe forms of spray dilution are shown in table 7.3.

Table 7.3

Terms and volume rates currently used in spray dilution ($l\ ha^{-1}$)

Terms	*Volume rates*	
	Ground crops	*Bushes and trees*
High volume (HV)	> 600	> 1000
Medium volume (MV)	200–600	500–1000
Low volume (LV)	50–200	200–500
Very low volume (VLV)	5–50	50–200
Ultra low volume (ULV)	< 5	< 50

Selection of the volume rate at which a chemical is applied tends to be left very much to the user's discretion with recommendations ranging over several hundred litres per hectare and generally determined by the output characteristics of a particular sprayer. High-volume application involves spraying the target to run-off with the aim of covering all plant surfaces. Use of lower volumes applied as a spray containing a large number of smaller discrete droplets produces at least as good control. If volume is reduced and the same level of chemical application is required it is necessary to increase concentration and to use a finer *droplet spectrum* (this term is used to describe the distribution of spray droplets into size groups). The change from high-volume to low-volume spraying necessitates use of the terms *coverage* and *dose*. In the past dose has related the amount of active ingredient applied to unit area of land; it now more properly relates to unit area of plant surface and is expressed in $\mu g\ cm^{-2}$. Coverage is applied to the proportion of the total area treated with spray. Cover may vary from a complete film on the target to a distribution of discrete droplets. Particularly with systemic fungicides (section 6.3.3) it is unnecessary to achieve complete coverage. Even with non-systemic materials complete coverage is not necessary since each speck of deposited chemical has a zone of fungicidal influence (Courshee, 1967). Medium-volume and low-volume spraying, which involve applying the same amounts of chemical but with less water, became feasible for orchard crops with the advent of air-carrier machines and for ground crops with more efficient hydraulic sprayers using smaller jets. Since low-volume sprays only partly cover a target there may be effects from redistribution of even non-systemic materials. Water on the foliage from rain or dew has been shown to redistribute Bordeaux and Burgundy mixtures, copper oxychloride and zineb (Hislop and Cox, 1970). Redistribution was limited to the channels between veins and consequently varied with leaf structure, wettability and the different forms of leaf trichome. Some chemicals will reach the target organism after application to the plant by redistribution in the vapour phase.

Application volumes have been reduced still further with the introduction of machines using very low and ultra low volume rates. The latter has been defined as the minimum volume applied per unit area required to achieve economic control (Anon, 1971). This can only be achieved by control of droplet sizes and densities on the target foliage. It is now being accepted that an effective way of increasing crop cover using a given volume of spray is by reduction in droplet size (Hislop and Baines, 1980). High-volume sprays apply droplets in excess of 200 μm diameter while low-volume sprays consist of droplets in the range 50–200 μm (Fulton, 1965). This may be placed in perspective by comparison with the size of droplets in various forms of natural precipitation as shown in table 7.4.

Table 7.4

Droplet size of various types of natural precipitation (after Ripper, 1955)

Droplet diameter (μm)	*Precipitation*
> 3000	Cloud burst
2000–3000	Heavy rain
1000–2000	Modest rain
500–1000	Light rain
200–500	Drizzle
100–200	Mist
30–100	Wet fog
5–30	Dry fog
0.5–5	Smoke

From table 7.4 it is apparent that high-volume application equates with drizzle and low volume with mist. Most sprayers will produce a range of droplet sizes within these wide limits. Where droplet size is known and can be held constant and droplet density can be estimated, then the spray volume per unit area may be calculated. For instance, if a spray of 100 μm diameter droplets is applied at a density of 50 droplets cm^{-2} then the minimum volume required is 2.6 l ha^{-1}. Minimum volumes of spray required to achieve a droplet pattern with a density of one droplet mm^{-2} with droplets of different diameters are shown in table 7.5.

Table 7.5

Spray volumes and droplet sizes required to achieve 1 droplet mm^{-2} (Bals, 1975a)

Droplet diameter (μm)	*Spray volume ($l\ ha^{-1}$)*
10	0.005
20	0.042
30	0.141
40	0.335
50	0.655
60	1.131
70	1.797
80	2.682
90	3.818
100	5.238
200	41.905
500	654.687

A droplet size classification has been suggested by Matthews (1975) and is outlined in table 7.6.

Table 7.6

Droplet size classification (after Matthews, 1975)

Droplet volume median diameter (μm)	*Droplet classification*
< 50	Aerosol
51–100	Mist
101–200	Fine spray
201–400	Medium spray
> 400	Coarse spray

Droplet size has most frequently been measured by *Volume Median Diameter (VMD)*, where volume is used as the parameter to quantify droplet size by dividing the volume of spray droplets about their mean. A few large droplets may account, however, for a major proportion of the total volume, thereby increasing the VMD value. On its own, therefore, this does not properly describe the range of droplet sizes and so use is made of the *Number Median Diameter (NMD)*, which is the mean diameter of droplets without reference to volume and emphasises the number of smaller droplets. The ratio (*r* value) VMD:NMD indicates the range of droplet sizes and in practice the applied spray droplet size will be more uniform the nearer this ratio approaches unity.

A major difficulty with reducing droplet size is that this increases the risk of environmental contamination by drift. Small droplets of less than 50 μm diameter take very significant times to fall under gravity and consequently are exposed to air currents for long periods, leading to greater risks of drift. This effect is minimal with droplets larger than 200 μm diameter. Use of small droplets means that a large proportion of the spray volume may fail to be collected within the target crop (Bache and Sayer, 1975). These authors found that peak deposition of droplets less than 50 μm diameter downwind was proportional to the height of the nozzle and the intensity of air turbulence. Large droplets were relatively unaffected by air turbulence. There are also considerable evaporative effects on spray droplets smaller than 50 μm diameter. Even in temperate conditions of 20°C and 80 per cent RH, droplets of this size evaporate in less than 50 s leaving only non-volatile material. In tropical conditions this process is greatly accelerated.

Hydraulic nozzles, although simple in design, produce a spray jet that is governed by many variable factors. Droplets in a wide size range and having different velocities are formed by hydraulic nozzles. Target plants may be exposed to a few large high-velocity droplets and many smaller slower ones. Attempts to reduce some of the variability in spray application have led to the development of the concept of *Controlled Droplet Application (CDA)*. This aims to reduce drift and to permit better spray application, timing and logistics. This is achieved by the use of appropriate droplet sizes and densities for a particular target. CDA stresses the need for a defined narrow droplet spectrum, although this need not necessarily mean small droplets. The basic aims of CDA were defined by Linke (1978) as follows.

1. *Elimination of small drift-prone droplets.*
2. *Reduced drift hazard*, to increase the number of available spraying days and to improve the chances of better timing of chemical applications.
3. *Enhanced and more predictable results.*

4. *Use of smaller volume rates*, to improve the general logistics of a spraying operation.
5. *Reduced weight of equipment*, to cause less damage to soil structure and crop foliage.

This is accepted to mean the application of suitable formulations of chemicals in very low volumes and with a defined and restricted droplet size range required to give acceptable biological results and economic response.

The use of correctly sized uniform droplets is emphasised by CDA to apply minimal doses of chemical and is a logical extension of low-volume spraying. Wettable powder and emulsifiable concentrate formulations (section 6.4.1) may be applied by CDA but if evaporation has to be reduced with droplets of less than 100 μm diameter, or an adjuvant is needed to enhance redistribution or cuticle penetration, then special formulations are required. The most successful method so far of controlling droplet size is by use of centrifugal energy nozzles, that is, by spinning discs or cups. Droplet size is determined by the rotational speed of the disc or cup. Liquid is fed from a reservoir tank to near the centre of a spinning disc or cup and is then forced by centrifugal processes towards the edge of the spinning surface. From there it is thrown off either as single droplets or at high flow rates as ligaments. These usually break up into two distinct droplet sizes—the main droplets and smaller satellites. If the flow rate is too high, sheets of liquid leave the spinning disc and break up at random in a manner similar to sprays from a hydraulic nozzle. Control of both flow rate and disc speed is essential to ensure the correct droplet size. Droplet formation is improved by the use of toothed or grooved discs.

The types of CDA sprayer developed so far include the following.

1. *Hand-carried battery-operated spinning disc sprayers*. These are lightweight machines with a plastic spray head and small D.C. motor which drives the rotating disc. Other parts include a liquid reservoir, usually a screw-on bottle, and a handle which contains the battery power supply.
2. *Hand-carried motorised spinning disc machines*. The discs are mounted in front of a fan which provides a directional airstream for use in glasshouses where natural air movement is insufficient to disperse the spray droplets. Power may be supplied by battery, A.C. electricity from a mains source or portable generator or from a small two-stroke engine incorporated with the sprayer.
3. *Tractor-mounted sprayers*. Centrifugal energy nozzles have been used for herbicides and may be extended to fungicide application. Orchard experiments with fungicide application have used modified axial-fan sprayers with centrifugal energy nozzles mounted in place of hydraulic

nozzles. These have been used in the USA for control of *Venturia inaequalis* (apple scab—section 2.6.5) (Mathee *et al.*, 1975) and commercial models are becoming available in the UK (Lawson, 1982). There is regular use of CDA sprayers for glasshouse vegetables in the UK (Sylvester, 1978) and some experiments have been made with glasshouse ornamentals (Jarrett *et al.*, 1978).

The initial concept behind CDA was to use 'liquid dusting' of fungicides with 70-90 μm droplets. In practice this has progressed to 'liquid granules' with droplets of 280-300 μm since drift is minimal with droplets in this size range and yet they are sufficiently small to give effective cover with volumes down to 10 l ha^{-1}. Reduced voltages produce lower atomiser speeds and thereby larger droplet sizes but the atomiser speed must be constant to produce a constant droplet size. For sizes of 280-300 μm with 80 mm discs, rotational speeds of 1500-2000 rpm are needed and the droplets are thrown off the disc at 30 kmph in a horizontal plane. These will be unaffected by air turbulence of 5-10 kmph and, since the sprayer is held close to a target, the droplet terminal velocity of 1 m s^{-1} ensures that they reach the target before wind can interfere with their initial momentum (Bals, 1975b).

At present the adoption of this technique for crop protection is slow, except perhaps in the glasshouse sector for the following reasons.

1. *Unsuitable formulation*—not all agrochemicals are suitable for formulation in highly concentrated liquid form, consequently a plurality of systems is needed to suit a range of sprays.
2. *Development for conventional spraying*—there are large numbers of chemicals that have been researched, formulated, cleared and marketed for conventional low-volume application. The cost and time involved in repeating this process for any new application method is immense.
3. *Insufficient development of sprayers*—the sprayers so far marketed have achieved light weight and low cost but in some cases at the expense of robustness. They have not offered sufficient control of the small droplets they produce to provide reliable efficacy and environmental advantages which are available in theory.
4. *Unsuitable vehicles*—a range of versatile lightweight and fast spraying vehicles still needs to be developed to provide large-scale growers with the advantages of ultra-low volume spraying (Harris, 1981).

At present CDA may have few advantages where low-volume conventional spraying can be used when drift is no problem and in practice growers may be reluctant to invest in expensive equipment that is novel and possibly of unreliable effect (Taylor, 1981).

7.2.5 Electrodynamic application

Conventional hydraulic sprayers apply droplets that each contain far more chemical than is required to kill or inhibit a pathogen. This also applies to CDA sprayers producing droplets of 250 μm. If droplet size is reduced to 40 μm there is still sufficient active chemical per droplet and it is possible to greatly increase the number of droplets, that is, the number of active sites per unit area of target, or to reduce still further the volume of liquid spray required. Spraying droplets of this size, however, raises the inevitable problems of drift and evaporation prior to impact at the target (section 7.2.3). A potential solution to this problem is to charge the spray droplets with opposite polarity to the target, thereby causing physical attraction between them. As plant surfaces are negatively charged (Moore, W., 1921) the spray must be positively charged. Electrostatic crop dusting began to be advocated in the 1940s (Hampe, 1947) but machines were unwieldy and unreliable. Development of miniature, solid-state, lightweight generators that produce high voltages makes it feasible to impart charges to sprays under field conditions. This permits application rates to be reduced to 0.5–15 l ha^{-1}. So far three forms of sprayer are in the process of development.

1. *North American designs in which the spray nozzle is embedded in an electrode.* By using the air-carrier technique, spray droplets 30–50 μm diameter are forced through an intense electrical field and then propelled towards a crop by the air stream that has been used to atomise the spray liquid.
2. *A spinning-disc machine developed by Sheffield University and Rothamsted Experimental Station, UK.* In this an electrode carrying 30 kV charges spray particles as they are broken up by the spinning disc.
3. *A technique developed by Imperial Chemical Industries, UK, which requires no mechanical energy to form a spray of fine droplets.* Liquid is passed between concentric electrodes and subjected to an intensely divergent electrical field. This forms and charges the droplets with the same polarity as the nozzle electrode and electrical repulsion drives the droplets away from the nozzle at high velocity. The constant force-field applied to each droplet during transit to the target minimises any effects resulting from extraneous forces such as gravity or air movement and leads to accurate targeting and low drift. This technique is highly efficient in terms of energy used; the machine can operate for up to 60 h using four 1.5 V batteries and has no moving parts.

The advantages of these machines are increased number of droplets, uniform distribution on all plant surfaces because of mutual droplet repulsion and significant reduction in overspray, run-off and drift (Coffee, 1981).

It is thought that electrostatic spraying will evolve as a technique to complement but probably not to replace other methods. It could be envisaged that the user will be supplied with an integrated package containing nozzle plus chemical in a sealed unit. Consequently there would be no nozzle calibration or maintenance and no filling or mixing of chemicals. The dose rate (Frick, 1970), flow rate and droplet size would be predetermined for optimal effect and there would be no toxic hazards resulting from spillage in mixing and filling nor the possibility of wrong mixing.

7.2.6 Electronic control of sprayers

The application of a constant volume of spray per unit surface area by conventional hydraulic and air-carrier machines is moderated by the displacement of the equipment and flow rate at the nozzles. This volume should remain constant irrespective of topography, soil conditions and poor adherence by the sprayer. Monitoring and control equipment is beginning to appear which electronically compares, by means of microprocessors, the actual performance of a sprayer with pre-set optimal levels. This is achieved by use of tachometers and manometers to measure speed and volume, and the data from them are compared electronically with ideal requirements (figure 7.11).

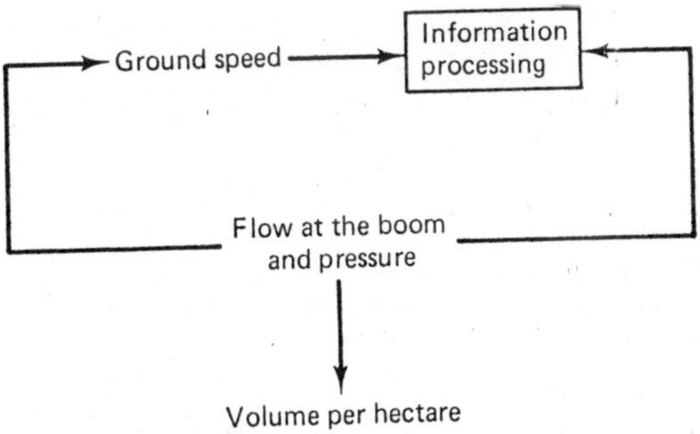

Figure 7.11 Diagram of electronic monitoring of the spraying operation

At present this equipment largely monitors the spraying operation and indicates to an operator either visually or audibly when gross variations occur although solenoid valves can be incorporated to correct increases in pressure. It is possible to envisage future equipment that will be capable of correcting most faults. A further refinement suggested is the monitoring of machine speed by radar which could eliminate errors resulting from slippage and spinning wheels (Givelet, 1981; Lawson, 1982).

The general trends in spray application technology are as follows.

1. *Improvement of operational efficiency*, by raising the percentage of time spent applying chemical by use of wider booms, faster vehicles, low ground pressure vehicles for use in wetter conditions and reduced spray volumes.
2. *Optimisation of chemical usage*, particularly in the application and placement of materials with changes in physico-chemical forms used, particle sizes and position in relation to the target organisms in order to produce better results technically, economically and ecologically.
3. *Safety and environmental care*, by reducing the gross amount of chemical used to produce less contamination of non-target crops and land areas by drift, run-off and leaching with fewer hazards to operators particularly by the use of closed systems for handling and metering (Darter, 1981).

The trend in chemical materials is for fewer new ones to be developed because of the high costs of research and development particularly in obtaining toxicological data for regulatory purposes. There is a general reluctance to apply for clearance for uses with many horticultural crops which are considered 'minor uses', where the potential market does not justify the investment costs involved. This is a particularly serious problem in view of the very high value of many individual horticultural plants and products.

REFERENCES

ANON (1971). Application and dispersal of pesticides. *Technical Report Series World Health Organisation.* No. 465, World Health Organization, Geneva

ANON (1980). *Guidelines for Applying Crop Protection Chemicals.* Booklet No. 2272. Ministry of Agriculture, Fisheries and Food, London

BACHE, D. H. and SAYER, W. J. (1975). Transport of aerial spray. I. A model of aerial dispersion. *Agricultural Meteorology*, 15, 257–71

BALS, E. J. (1975a). The importance of controlled droplet application (CDA) in pesticide applications. *Proceedings 8th British Insecticide Fungicide Conference*, 1, 153–60

BALS, E. J. (1975b). Development of a cda herbicide handsprayer. *Pest. Articles News Summary*, 21, 345–9

BRUNSKILL, R. T. (1956). Factors affecting the retention of spray droplets on leaves. *Proceedings 3rd British Insecticide Fungicide Conference*, 2, 593–603

BURCHFIELD, H. P. (1959). Physical chemistry of fungicidal action: physical properties and chemical reactivities in relation to the effectiveness of fungicides. 293–303, in *Plant Pathology–Problems and Progress 1908–1958* (eds C. S. Holton, G. W. Fischer, R. W. Fulton, H. Hart and S. E. A. McCallan), Wisconsin University Press, Madison

COFFEE, R. A. (1981). Electrodynamic crop spraying. *Outlook on Agriculture*, 10, 350–6

COLE, I. (1978). *Efficient soft fruit spraying*, Ministry of Agriculture, Fisheries and Food, Agricultural Development and Advisory Service, Chelmsford, Essex

COURSHEE, R. J. (1967). Application and use of foliar fungicides. 239–86, in *Fungicides—An Advanced Treatise*, Vol. I, *Agricultural and Industrial Applications Environmental Interactions* (ed. D. C. Torgeson), Academic Press, New York

DARTER, I. E. (1981). Trends in application technology. *Outlook on Agriculture*, 10, 319–20

EVANS, E. (1968). *Plant Diseases and their Chemical Control*, Blackwell, Oxford

FRICK, E. L. (1970). The effects of volume, drop size and concentration and their interaction on the control of apple powdery mildew by dinocap. 23–33, in *Symposium for Research Workers on Pesticide Application*, British Crop Protection Council Monograph No. 2

FULTON, R. H. (1965). Low volume spraying, *Annual Review Phytopathology*, 3, 175–96

GIVELET, M. P. (1981). Electronic control systems in pesticide application machinery. *Outlook on Agriculture*, 10, 357–60

HAMPE, P. (1947). Les poudreuses electriques à champ ionisé. *Revue de Viticulture*, 93, 259–61

HARRIS, D. A. (1981). Development of ULV techniques. 175–81, in *Proceedings of Crop Protection in Northern Britain 1981 Conference* (available from Scottish Crop Research Institute, Dundee)

HISLOP, E. C. and BAINES, C. R. (1980). An analysis of some spray factors affecting the protection of foliage by fungicides. 23–33, in *Spraying Systems for the 1980s*, British Crop Protection Council Monograph No. 24

HISLOP, E. C. and COX, T. W. (1970). Local re-distribution of fungicides on leaves by water. *Annals Applied Biology*, 66, 89–101

JARRETT, P., BURGES, H. D. and MATTHEWS, G. A. (1978). Penetration of controlled drop spray of *Bacillus thuringiensis* into chrysanthemum beds compared with high volume spray and thermal fog. 75–81, in *Controlled Drop Application*, British Crop Protection Council Monograph No. 22

KEITT, G. W. and JONES, L. K. (1926). Studies of the epidemiology and control of apple scab. *Wisconsin Agricultural Experiment Station Research Bulletin*, 73

LAPWOOD, D. H. (1961). Potato haulm resistance to *Phytophthora infestans*. III. Lesion distribution and leaf destruction. *Annals Applied Biology*, 49, 704–16

LAWSON, G. (1982). Cutting spraying costs. *The Grower*, August 12, 45–7

LINKE, W. (1978). CDA—a review of developments to date. *Proceedings 14th British Weed Control Conference*, 3, 1047–57

LODEMAN, E. G. (1910). *The Spraying of Plants*, Macmillan, New York

MARTIN, H. and WORTHING, C. R. (eds), (1976). *Insecticide and Fungicide Handbook*, Blackwell, London

MATHEE, F. N., THOMAS, A. C., SCHWABE, W. F. S. and NEL, E. W. (1975). Low volume and ultra low volume sprays to control apple scab (*Venturia inaequalis*). *The Deciduous Fruit Grower*, 25, 183–6

MATTHEWS, G. A. (1975). A graticule for classification of spray droplets. *Pest. Articles News Summary*, 21, 343–4

MATTHEWS, G. A. (1981). Developments in pesticide application for the small scale farmer in the tropics. *Outlook on Agriculture*, 10, 345–9

MOORE, M. H. (1964). Glasshouse experiments on apple scab. I. Foliage infection in relation to wet and dry periods. *Annals Applied Biology*, 53, 423–35

MOORE, W. (1921). Spreading and adherence of arsenical sprays. *Technical Bulletin* 2, University of Minnesota Agricultural Experiment Station

MORGAN, N. G. (1981). Minimising pesticide waste in orchard spraying. *Outlook on Agriculture*, 10, 342–4

RIPPER, W. E. (1955). Application methods for crop protection chemicals. *Annals Applied Biology*, 42, 288–324

SYLVESTER, N. K. (1978). The relevance of ultra low volume spraying for the application of fungicides to protect winter lettuce under glass against grey mould. 231–41, in *Controlled Drop Application*, British Crop Protection Council Monograph No. 22

TAYLOR, W. A. (1981). Controlled drop application of herbicides. *Outlook on Agriculture*, 10, 333–6

FURTHER READING

ALLEN, J. G., BUTT, D. J., DICKER, G. H. L. and HUNTER, L. D. (1978). A comparison of the efficiency of sprays of two drop-size ranges in an apple orchard. *Pesticide Science*, 9, 545–54

ANON (1958). *The Comparison of High- and Low-Volume Spraying Techniques on Fruit and Ground Crops*. Symposium Monograph No. 2, Society Chemical Industry, London

ANON (1965). *Field Crop Sprayers*. Mechanisation Leaflet No. 2, Ministry of Agriculture, Fisheries and Food, London

ANON (1970). *Orchard Sprayers*. Mechanisation Leaflet No. 28, Ministry of Agriculture, Fisheries and Food, London

ANON (1978). *Controlled Drop Application*, Monograph No. 22, British Crop Protection Council, Croydon

ANON (1979). *Horticultural Sprayers for Small Areas*. Booklet 2070, Ministry of Agriculture, Fisheries and Food, London

ANON (1980). *Guidelines for Applying Crop Protection Chemicals*. Booklet no. 2272, Ministry of Agriculture, Fisheries and Food, London

ANON (1981). *Controlled Droplet Application of Agricultural Chemicals*. Leaflet 792, Ministry of Agriculture, Fisheries and Food, London

ANON (1982). *British Agrochemicals Association Ltd Manual* (looseleaf). British Agrochemical Association Ltd, London

BRANDES, G. A. (1971). Advances in fungicide utilization. *Annual Review Phytopathology*, 9, 363–86

BYASS, J. B., and LAWRENCE, D. C. (1976). The scope for the use of controlled drop application (CDA) on large cereal growing farms. *Proceedings 13th British Crop Protection Conference, Weeds*, 2, 363–68

CHAMBERS, D. (1981). A look at electrostatic spraying. *Agrospray*, 4, 16–17

FARMERY, H. (1975). Controlled droplet engineering—an engineer's viewpoint. *Proceedings 8th British Insecticide Fungicide Conference*, 1, 171–74

GRIFFIN, M. F. and THOMAS, B. (1979). The safe use of chemicals in agriculture. *Chemistry in Agriculture*, Chemical Society special publication No. 36, 237–50

HAWKER, M. J. F. and KEENLYSIDE, J. F. (1977). *Horticultural Machinery*, Longman, London

MATTHEWS, G. A. (1975). Determination of droplet size. *Pesticide Abstracts News Summary*, 21, 213–25

MATTHEWS, G. A. (1979). *Pesticide Application Methods*, Longman, London

MORGAN, N. G. (1973). Effective spray application techniques. *Annual Report Long Ashton Research Station for 1972*, 103–6

MORGAN, N. G. (1974). Development of spraying techniques. *Annual Report Long Ashton Research Station for 1973*, 105–8

MORGAN, N. G. (1976). Spray application research at Long Ashton. *Agricultural Research Council Research Review*, 2, 84–5

ROBERTSON, J. (1978). *Mechanising Vegetable Production*, Farming Press, Ipswich

RUTHERFORD, I. (1977). The faults, the utilisation and performance of field crop sprayers. *The Agricultural Engineer*, 32(2), 40–4

RUTHERFORD, I. (1977). ADAS views of impending farm developments. *National Institute of Agriculture Engineering Subject Day Paper* No. 6, published in *N.I.A.E. Report* No. 25

SIMPER, H. (1977). *Farmers Guide to Arable Machinery*, Farming Press, Ipswich

TAYLOR, W. A., MERRITT, C. R. and DRINKWATER, J. A. (1976). An experimental tractor mounted, very low volume, uniform-drop-size sprayer. *Weed Research*, 6, 203–8

TEMPLE, K. B. (1980). *An Introduction to Pesticides*, Shell Chemicals Ltd, Ely

TORGESON, D. C. (ed.) (1967). *Fungicides—An Advanced Treatise*, Vol. I. *Agricultural and Industrial Applications, Environmental Interactions.* Academic Press, New York

WALKER, J. O. (ed.) (1980). *Spraying Systems for the 1980s.* Monograph No. 24, British Crop Protection Council

8 CONTROL: PLANT HEALTH REGULATION

National and international trade in horticultural produce and planting material carries with it the risks of disseminating pathogens. Control of these risks by using plant health regulation is attempted by domestic legislation and international co-operation. The latter is fostered under an International Plant Protection Convention of 1951. Three forms of plant health regulation are described in this chapter, which although they are treated separately, are in practice closely interrelated.

8.1 PHYTOSANITARY INSPECTIONS AND QUARANTINE

Quarantine may be defined as all operations associated with preventing the dissemination of unwanted organisms within and between territories. This definition embraces the issue of phytosanitary certificates. Quarantine should prevent the spread of pathogens but should not seek to restrict product movement between areas where the same pathogens are known to occur except if an eradication programme is operating. Regulations should not be established to improve the competitive position of indigenous produce. The economic, social and environmental benefits to be gained from the application of procedures should outweigh their administrative costs and the costs to affected industries and the public. To have a reasonable chance of success there must be co-operation between governmental agencies and the public. To this end an extensive knowledge of the biology of the organism is essential so that simple and sensible procedures can be framed. Quarantine is best applied at the source of spread rather than at its destination. As conditions alter or more facts become available, the quarantine measures need modification either to add restrictions or to remove them.

Most countries operate some form of phytosanitary inspection service that attempts to prevent importation of new pathogens and to provide evidence in the form of phytosanitary certificates indicating that plant material for export is free from pathogens. This is based on the International Plant Protection Convention under which an exporting country is obliged to inspect and provide certificates in conformity with the requirements of an importing country. These schemes are closely co-ordinated by eight world regional plant protection organisations which attempt to prevent the spread of pathogens on or in plants and products that are traded between countries. The plant protection organisations also disseminate information on methods of plant protection and attempt to simplify and unify phytosanitary regulations and the inspection requirements of member countries.

Importation restrictions for plant health reasons are governed by consideration of which pathogenic species warrant action, the potential ability of a pathogen to survive in the importing country and how prevention may be achieved in a cost-effective manner. The likely risks from a non-indigenous pathogen may be evaluated from the following.

1. *Its ability to spread and cause damage over wide areas or to impair the economic viability of specialist crops in the new location.*
2. *Its capacity to lower the capital value of land through its persistence in soil.*
3. *The availability of control measures and its potential for adding considerably to disease control costs.*
4. *Its capacity to undermine the work of plant breeders.*
5. *Its capacity to restrict subsequent trade by creating difficulties in meeting the phytosanitary requirements of other countries.*
6. *Its capacity to mask other quarantine subjects.*
7. *Its properties that interfere with the quality of indigenous produce.*

Upon the arrival of planting material, an importing country may use a number of methods to ensure conformity with its needs. These increase in severity from a system of check inspections, to inspection of all planting material of specified types, to simple restrictions on initial growing or distribution (known as open post-entry quarantine) to restricted entry via a special quarantine facility where the material is grown, observed and subjected to tests.

The successful use of import restrictions to prevent entry of non-indigenous pathogens depends on the degree of enforcement and whether or not other avenues for entry exist. It may only be necessary to prohibit importation during certain seasonal risk periods, or of soil adhering to plants, or of rooted plants or of various types of packaging material. The problem may be overcome by chemical treatment such as fumigation in the exporting country. To prevent entry of some pathogens, growing-season inspections in the country of origin may be required prior to phytosanitary certification before export; this necessitates an evaluation of the value of such inspections. Where risks of importing particular pathogens to the industries of a country are very high then that country may require that its own inspectors carry out pre-export inspections within the exporting country; for example, USA inspectors have conducted surveys in Japan to prevent entry of *Xanthomonas citri* (citrus canker) into American crops. The value of all inspections, however, depends on the sampling techniques being capable of identifying infected material within a consignment and these are not always successful. To be accurate many samples need to be drawn in a random manner. Inspection methods should relate the detection methods required with appropriate statistically valid sample numbers. Pre-entry requirements in one country before export to another may require absence of a pathogen from the place of production for a specified period, a radius of freedom in the production area or maintenance of stringent control by chemical and husbandry methods or derivation from tested (for example, virus-tested) material.

8.2 CERTIFICATION SCHEMES

8.2.1 Vegetative planting material

By virtue of vegetative propagation techniques, plant stocks accumulate pathogens, especially viruses and virus-like organisms, thereby declining in vigour and productivity. Certification seeks to remedy this by ensuring that a supply of healthy material of proven origin is available. The objective is to obtain a nucleus of healthy material that may be multiplied under specified conditions, checked and given a certificate. This requires growing-season inspections, in some instances laboratory tests and the establishment of tolerance levels for pathogen presence. The ideal pathogen for eradication by certification would have easily recognisable and distinct infection symptoms on the aerial vegetative organs and would be spread exclusively on or in planting materials; few if any pathogens have such distinct attributes.

Where pathogen spread is aerial (by insect vectors, wind-blown debris or pollen-borne) certification may control it by isolation, roguing, chemical control of vectors and deblossoming. Soil-borne pathogens may be combated by producing planting material in areas known to be pathogen free. Such freedom is established by laboratory testing of field-drawn soil and host samples as in the case of *Phytophthora fragariae* (strawberry red core). Where spread is by contact between host plants as with potato virus X, control may be achieved by clonal selection followed by the multiplication of pathogen-free mother plants, which are check-tested by serology, and hygiene standards are instituted when agricultural operations take place. The term 'virus free' (more correctly virus tested) is often applied to certified material. This means that the material has been tested (indexed) and is free from known viruses and virus-like agents following adequate use of present detection methods.

To operate successfully, certification schemes must be simple and thereby command the respect of farmers and growers. This may be exemplified by the scheme for strawberry, a crop with a long history of decline in specific commercially dominant cultivars because of virus infections that have subsequently been eradicated and whose health has been maintained by certification schemes using the route shown in figure 8.1. The figure shows that a maximum of four generations of vegetative reproduction are therefore permitted following the release of virus-indexed stock.

Virus-indexed mother plants (a nucleus produced by East Malling Research Station, Kent)

↓ Release of a small number of virus-tested runners

Foundation stocks produced by a Nuclear Stock Association with growers using closely specified and supervised conditions

↓

Elite stock for sale to plant raisers who need not necessarily be members of the Nuclear Stock Associations

↓ Entry to an official Certification Scheme that stipulates conditions of runner production, length of rotation, isolation, inspection and standards of health and purity

Specific (Special) stock (SS grade) → sold to fruit producer

↓ Sold to plant raiser

'A' grade stock (Standard (S) grade in Scotland)

↓

Sold to fruit producer

Figure 8.1 Route for the production of virus-tested strawberries

8.2.2 Seed certification

Seed testing ensures that germination and purity reach specified levels; in addition to obtain the full benefits of pathogen-resistant cultivars, seed certification schemes are essential in order to maintain trueness to type. This requires the establishment of quality standards for seed and for each seed batch to be traceable. This necessitates seed testing in the laboratory before field sowing and after harvest, crop inspections for isolation, freedom from pathogens and mixtures with other cultivars. Certification follows the seed breeders' material (the tiny quantities resulting from a breeding programme) through pre-basic and several further, but specified, numbers of generations for field production. Each generation is examined in the field and by laboratory tests to ensure that the prescribed standards are maintained. Some seed certification schemes are co-ordinated internationally to ensure that seed grown in one country may be sold freely in another, for example, within the European Community.

Two groups of pathogens may be controlled by certification.

1. *Those depending mostly or entirely on seed-borne inocula*, either in the seed tissues, attached to the outside of the testa or carried as dust with the seed. Here the use of pathogen-free ('clean') seed is an effective and cheap means of eradicating a pathogen almost completely. An example is lettuce mosaic virus (section 3.5.2).
2. *Pathogens that are widespread and expensive to control*, often developing only when environmental conditions are especially conducive. Resistant cultivars are a valuable means of containing such pathogens. Certification maintains the character of resistant cultivars, thereby prolonging their usefulness. Examples are tomato cultivars with resistance to *Verticillium albo-atrum* (section 2.6.9) and *Fusarium oxysporum* f. sp. *lycopersici* (vascular wilts).

8.3 ERADICATION AND CONTAINMENT

Eradication of an established pathogen has never been attempted in the UK but in the USA a sustained 30-year campaign to eradicate *Xanthomonas citri* (citrus bacterial canker) was successful. This pathogen has not been found in the USA since 1943 despite having previously been endemic in wild *Citrus* spp. in the Everglades and other uncultivated areas and in citrus orchards of seven Southern States. In the UK the campaign to prevent the establishment of *Puccinia horiana* (chrysanthemum white rust) by eradication of foci of infection emanating from imported planting material has been successful and cost beneficial so far since chemical controls are only partially effective and resistance unlikely to be of long-term value because of the development of physiological races. Within the UK, co-operative crop-management methods can be used to contain but not eradicate pathogens such as virus yellows in sugar beet and *Peronospora viciae* (pea downy mildew). They are particularly effective in sugar beet and vegetable crops grown under contract for 'quick-freeze' processing, for example beans, brassicas and peas, where there is widespread control of crop-husbandry techniques and of the cultivars grown.

To eradicate alien pests that become established requires detection and identification of infection foci, surveys of the dispersion of the pathogen and evaluation of the costs and benefits of removing the pathogen following arrival. To develop an eradication programme it is necessary to know the rate and mode of spread, association with plant material, means by which eradication may be achieved, the resources in terms of facilities and manpower needed and the likelihood of reintroduction once eradication has taken place. The UK programme to eradicate *Erwinia amylovora* (fire-blight) is an example of a campaign that appears to have been partially successful (Ebbels and King, 1979). In practice this has become a special example of containment. The pathogen is now endemic in wild hosts such as hawthorn (*Crataegus* spp.) in southern England but by elimination of susceptible cultivars the pear industry has avoided severe losses due to *E. amylovora.*

REFERENCE

EBBELS, D. L. and KING, J. E. (eds) (1979). *Plant Health, The Scientific Basis for Administrative Control of Plant Diseases and Pests*, Blackwell, London

FURTHER READING

EBBELS, D. L. (1979). A historical review of certification schemes for vegetatively propagated crops in England and Wales. *Agricultural Development and Advisory Service Quarterly Review*, 32, 21–58

HOLLINGS, M. (1965). Disease control through virus free stock. *Annual Review Phytopathology*, 3, 367–96

LEVY, B. G. (1948). Certification schemes for fruit plants. *Report East Malling Research Station for 1947*, 178–81

MATHYS, G. (1977). Society supported disease management activities. 363–80, in *Plant Disease—An Advanced Treatise*, Vol. I. *How Disease is Managed* (eds J. G. Horsfall and E. B. Cowling), Academic Press, New York

MATHYS, G. and BAKER, E. A. (1980). An appraisal of the effectiveness of quarantines. *Annual Review Phytopathology*, 18, 85–101

RICHARDSON, M. J. (1979). *An Annotated List of Seed-Borne Diseases.* Phytopathological Paper no. 23. Commonwealth Agricultural Bureaux, Farnham Royal

SHEPARD, J. F. and CLAFIN, L. E. (1975). Critical analyses of the principles of seed potato certification. *Annual Review Phytopathology*, 13, 271–93

9 CONTROL: CULTURAL AND BIOLOGICAL METHODS

The basic approach with cultural control is to invoke those aspects of husbandry that promote sound crop growth and inhibit or otherwise obstruct pathogenic growth. This should avoid, delay or lessen the impact of disease. Conversely those practices that operate in an opposite direction should be eliminated. More than any other form of control, cultural methods emphasise that the objective of horticulture is to produce fruitful, high-yielding crops rather than simply to control plant pathogens. These controls act largely in a preventive manner and are applied in advance of invasion. They do not lend themselves to control pathogens that occur in a sporadic manner. It should be firmly borne in mind that advocating adjustments to crop environment is a highly risky business where knowledge of the effects of environmental change on crop economics is imperfectly understood.

In this chapter bacteria and fungi are treated separately from viruses. This is because direct use of resistance and chemical control of viruses have largely been ineffectual so far, and controls are largely directed at preventing vectors from reaching the crop. In many cases this has been accomplished by cultural methods. Control of fungi and bacteria, on the other hand, has succeeded with direct use of resistance and chemical methods and a resultant neglect of cultural techniques. Now that the limitations of resistance and chemical methods are becoming more obvious, attention is turning to cultural methods. This is linked with a more widespread appreciation of biological methods of control which, with a few notable exceptions, have not been investigated in depth. Unified use of several control methods is now being attempted and is termed *integrated control*.

9.1 FUNGI AND BACTERIA

9.1.1 Environment

Siting of crops in relation to climatic conditions may have a bearing on subsequent damage by pathogens. Frost pockets should be avoided since frost *per se* will cause damage but there is a relationship between frost damage and pathogenic winter-kill especially with perennial plantation crops. The incidence of *Leptosphaeria coniothyrium* (raspberry cane blight) and *Dasyscypha wilkommi*, the cause of larch canker, has been related to winter-kill of their respective hosts. Infection by root-invading pathogens that have free-swimming motile stages can be associated with sites having periodically high water tables; examples are *Plasmodiophora brassicae* (clubroot—section 2.6.1) or *Phytophthora* spp. causing ink disease of *Castanea* spp. (chestnut). Control of pathogens with a soil-borne phase has been related to improvements in soil drainage; clubroot (*P. brassicae*) incidence fell from 92 per cent in waterlogged soil to 10 per cent in well-drained sites. Other pathogens where similar claims have been made are powdery scab (*Spongospora subterranea*) of potato, and root and collar rot of top and soft fruit (*Phytophthora cactorum*). The effects of poor drainage may be exacerbated by use of heavy machinery when in wet conditions wheelslip causes smearing in the furrow bottom, leading to the development of a plough-pan impeding water percolation still further. At the other extreme it is suggested that excessively dry sites predispose *Pinus* spp. to infection by *Diplodia pinea*, *Populus* spp. to *Dothichiza populea* and onions to *Sclerotium cepivorum* (white rot). This effect is presumably related to the stresses imposed on the host by water shortage.

Crops grown in areas with a moist microclimate by virtue of proximity to large water masses are at risk from pathogens favoured by high humidity; examples are downy mildew on beet (*Peronospora farinosa*) in Essex or on lettuce (*Bremia lactucae*—section 2.6.3) in the Tamar Valley, UK. Moist microclimates are encouraged within crops by close planting. High-density stands increase humidity and limit air circulation, stimulating infection by pathogens causing downy mildew and damping-off syndromes. A feature of crop production in many low-lying and damp areas is the use of ridges for planting in order to encourage air movement and drainage, especially in the juvenile phases of crop growth. In the propagation phase use of block-raising techniques has automatically led to better spacing of seedlings with concomitant reduction in the incidence of pathogens such as *Peronospora parasitica* (cabbage downy mildew).

Protected crops present good examples of a relationship between high humidity (above RH 93 per cent), cool temperatures (7–16°C) and high incidence of *Botrytis cinerea* (grey mould—section 2.6.4) because the germination of spores is optimal under these conditions. Fewer tomatoes become infected by *B. cinerea* when the plants are grown with a constant day and night regime of 20°C than if the regime fluctuates between 20°C day and 13°C night temperatures. Further reductions in infection occur if humidity is held below 95 per cent. Use of thermal screening to economise on fuel by reducing night-time radiation from glasshouses has probably resulted in significant problems due to infection by *B. cinerea* because of the high RH generated at night and the cooling effects resulting from withdrawing the screens early in the morning. There is a significant correlation between points where moisture condenses on foliage and fruits of tomato and subsequent *B. cinerea* incidence (Strijbosch, 1976). Even incidence of those pathogens usually favoured by dry conditions, such as *Sphaerotheca pannosa* (rose powdery mildew—section 2.6.6), may be reduced by maintenance of a dry glasshouse atmosphere. Rotting of nursery-stock cuttings when propagated by mist-bench techniques is reduced if the frequency of misting is lowered once rooting begins. Mist benches produce an imbalance in temperature because the plant roots are heated from beneath and the aerial parts are cooled by mist, and, in this stress situation, invasion by rot-causing fungi and bacteria is facilitated.

Irrigation has been demonstrated to affect the severity and forms of pathogen that infect crops. This may be reflected in the ability of foliar-invading pathogens to colonise leaves of plants weakened by excessive application of water to the soil. Such an effect was demonstrated with *Alternaria senecionis* (leaf spot) of florists' cineraria (*Senecio cruentus*) (table 9.1).

Table 9.1

Effect of soil moisture content on the leaf area of Senecio cruentus *invaded by* Alternaria senecionis *(after Cooper, 1956)*

Soil water content	*Mean diseased leaf area (cm²) in experiment number*				
	1	*2*	*3*	*4*	*5*
High (pots kept permanently wet)	8.6	15.4	11.3	18.3	31.9
Low (water given only to prevent wilting)	4.2	4.1	3.2	0.3	6.1

The manner in which water is applied may also affect infection. Tomato plots irrigated from overhead sprinklers were more severely infected by *Stemphylium botryosum* f. sp. *lycopersici* (leaf spot) and *Xanthomonas vesicatoria* (bacterial spot) than those where irrigation was applied in furrows, probably because both these pathogens are favoured by the high humidity generated by sprinkler irrigation. The level of infection was related to the length of time that high humidity persisted after irrigation had stopped. This in turn was affected by wind velocity and cloud cover (Rotem and Cohen, 1966). Similar effects have been reported where overhead irrigation increased disease severity with *Ustilago maydis* (smut) on sweet corn and *Pseudoperonospora cubensis* (downy mildew) on banana. One technique that has been advocated for control of *Bremia lactucae* (lettuce downy mildew–section 2.6.3) is that the peat blocks containing seedlings should be stood out on to land that has previously been brought to field capacity. This obviates the need for overhead irrigation during much of the growing phase of the crop and adversely affects pathogen development. An instance where irrigation has helped to control a pathogen of plantation crops is in the use of misting for frost protection of apple cv. Golden Delicious in the USA. Misting delayed blooming and resulted in lower infection by *Venturia inaequalis* (apple scab–section 2.6.5) as a result of decreased inoculum establishment because the ascospores were released while the trees were still dormant.

For control of water-borne pathogens such as *Phytophthora cinnamomi* in hardy nursery-stock beds, hypochlorite or chlorine is added to the irrigation water. Infection by this pathogen may also be influenced by the substrate on which the containers are placed since *Azalea* spp. were more severely infected when placed on polyethylene sheet which prevented drainage than if gravel, sawdust or pine bark were used.

Levels of acidity-alkalinity (pH) are well demonstrated to affect the growth of fungi and bacteria in artificial culture. It is often expected, therefore, that adjustments of soil pH can be used to control plant pathogens but the issue is rarely clear cut. *Plasmodiophora brassicae* (clubroot–section 2.6.1) is generally cited as a pathogen that may be controlled by increasing soil alkalinity. *P. brassicae* will, however, develop in acid and some alkaline conditions; severity increases when inoculum levels are high and soil moisture and temperature are favourable to development (Colhoun, 1953, 1961). Possibly the chemical constituents of materials used to raise pH are more important than shifts in acidity or alkalinity. Trials in Newfoundland, California, USA and England have shown that applications of 40 t ha^{-1} of lime significantly reduce clubroot and raise soil pH above 7.2. Raising pH has been shown to reduce wilt of musk melon caused by *Fusarium oxysporum* f. sp. *niveum*, whereas raising soil pH from 4.9 to 6.3 reduced fruit and foliar infections of tomato caused by *Botrytis cinerea* (grey mould–section 2.6.4); increases above this level led to a rise in symptom development. This result demonstrates a problem inherent in raising pH above the optimum for crop growth in that poor growth resulting from the effects of too high pH may render the host more susceptible to infection. Other host–parasite combinations where increased soil alkalinity has been reported to reduce disease are *Botrytis fabae* (chocolate spot of broad bean), *Helicobasidium purpureum* (violet root rot of asparagus) and *Phytophthora cinnamomi* (on *Eucalyptus marginata*.

Temperature manipulation may be used as a means of pathogen control because many micro-organisms have well-defined temperature optima for growth. *Urocystis cepulae* (onion smut) is absent in the warmer southern USA but is a persistent and damaging pathogen in cooler more northerly crops. Above 25°C there is reduced teleutospore germination and hyphal growth; additionally at 29°C the host grows rapidly and passes through the susceptible juvenile stage to a non-susceptible state following expansion of the primary leaves (Walker and Wellman, 1926). Low temperatures during the initial stages of germination of bean, cucumber, pea and maize increase the liability of infection by *Thanatephorus cucumeris* (a cause of damping-off). It is suggested that this is due to reduced host vigour at low temperatures.

Artificially high temperatures may be used as a means of pathogen eradication. Work in Israel demonstrated that covering the soil surface with polyethylene sheet and the subsequent solar heating for 14 days will eliminate *Verticillium dahliae* (section 2.6.9) and reduce the inoculum levels of *Fusarium oxysporum* f. sp. *lycopersici* at depths of 0–25 cm (Katan, 1981). Similar work in the UK demonstrated this as a technique for combating *Plasmodiophora brassicae* (clubroot–section 2.6.1) on small areas of land.

Manipulation of glasshouse temperature can be used to combat some pathogens; for instance, temperatures 10–14.5°C may be used to control *Uromyces dianthi* (carnation rust). The 'quiet-growing system' for lettuce of 10°C day and 5°C night temperatures with use of pipe heat at 20°C for short periods to reduce RH prior to nightfall was a successful means of combating *Bremia lactucae* (lettuce downy mildew–section 2.6.3) in the Westland area of the Netherlands. Day temperatures are now reduced to 7°C and night temperatures to 2°C to save fuel.

Treatment of seed by soaking in hot water has been used for many years to eradicate seed-borne pathogens. Soaking in water at 50°C for periods up to 30 minutes has been used to control such pathogens as *Alternaria brassicicola* (brassica dark leaf and pod spot), *Leptosphaeria maculans* (brassica root rot and canker), *Septoria apiicola* (celery leaf spot) and *Xanthomonas campestris* (brassica black rot). This method is now largely superseded by the safer technique of soaking seed for 24 h in thiram solutions at temperatures of up to 30°C (Maude, 1970). *Pseudomonas lachrymans*, the cause of cucumber angular leaf spot, was controlled in Japan by treating seed with dry air at 70°C for 3 days. Hot water is used to control nematode infection of *Narcissus* bulbs and the addition of formaldehyde to the water will control *Fusarium oxysporum* f. sp. *narcissi* (basal rot).

Since pathogens rely on the host as a nutrient source, the state of host nutrition is likely to be a means by which diseases may be controlled. This has been an area of much research but with few unequivocal conclusions. Use of high levels of nitrogen fertiliser has in many cases been correlated with increased disease severity but it is well recognised that the form of nitrogen used is at least as important as the rate of application. In general, ammonium nitrogen tends to be associated with increased disease severity when compared with nitrate nitrogen, but this is not always the case (Huber and Watson, 1974). *Thanatephorus cucumeris* (causing lettuce damping-off) increased with the use of ammonium nitrogen and application of a range of rates of nitrate nitrogen had no effect on disease severity (Das and Western, 1959; Shephard and Wood, 1963). This has been correlated with increases of the amides, asparagine and glutamine, in ammonium nitrogen-fertilised plants (Weinhold *et al.*, 1972). Root rot and cortical diseases caused by fungi such as *Aphanomyces, Armillaria* (*Armillariella*), *Cercospora*, *Fusarium* and *Poria* are reduced by nitrate nitrogen and increased by ammonium nitrogen, but opposite results were obtained with *Diplodia*, *Gaeumannomyces*, *Pythium* and *Streptomyces*. These effects appear to operate on the soil microflora or root physiology rather than on the plant as a whole. Interplanting red alder, which can fix its own nitrogen, with conifers has been proposed as a means of control for *Poria* and *Armillaria* spp., which cause root rots (section 2.6.7). Soil under a mixed stand of red alder and conifers contains more nitrate nitrogen than adjacent soil under pure conifers. Since *P. weirii* and *A. mellea* do not utilise nitrate nitrogen but grow well with ammonium nitrogen or amino nitrogen, red alder has the potential for a form of cultural control of these soil-borne pathogens by stimulating nitrification and increasing the soil nitrate nitrogen content. It has been suggested that the mechanism for this form of control results from an inability by these pathogens to utilise nitrate nitrogen for growth because of a lack of nitrate reductase enzyme and a stimulation of antagonistic *Streptomyces* spp. which readily utilise nitrate nitrogen (Li *et al.*, 1967).

Vascular wilts of potato and tomato caused by *Verticillium albo-atrum* (section 2.6.9) are stimulated by nitrate nitrogen (Walker *et al.*, 1954) and, in the case of potatoes, symptom severity is reduced by ammonium nitrogen. Other pathogens causing vascular wilts such as *Fusarium oxysporum* f. sp. *vasinfectum* (cotton wilt), *F. oxysporum* f. sp. *conglutinans* (cabbage yellows), *F. oxysporum* f. sp. *lycopersici* (tomato wilt) and *F. oxysporum* f. sp. *pisi* race 1 (pea wilt) were reduced in severity as levels of nitrate nitrogen increased (Walker, 1946; Gallegly and Walker, 1949). This is despite fusarial-wilt pathogens being primarily xylem-invading and able to utilise nitrate nitrogen. Bacterial pathogens causing wilt and canker of tomato (*Pseudomonas solanacearum* and *Corynebacterium michiganense*) increased in severity concomitantly with increased levels of nitrate nitrogen (Gallegly and Walker, 1949; Walker *et al.*, 1954).

There is a similarly confused picture with foliar-invading pathogens. Rust and powdery mildew pathogens are in general stimulated by nitrate nitrogen and retarded by ammonium nitrogen (Daly, 1949; Last, 1953). Pathogens such as *Botrytis fabae*, causing chocolate spot on broad bean, react in the opposite manner (Blakeman, 1981; Sol, 1967). Similarly, diseases of turf caused by *Phialophora cinerescens* and *Fusarium culmorum* are more severe with high rates of nitrate nitrogen fertiliser (Nilsson and Nelson, 1964) whereas *Sclerotinia homoeocarpa* (dollar spot) and *Thanetephorus cucumeris* (causing brown patch) are discouraged by this form of nitrogen (Couch and Bloom, 1960; Endo, 1966; Bloom and Couch, 1958).

In general, nitrogen form affects characters such as host-resistance expression, plant-growth constituents, exudates, soil and rhizosphere pH and pathogen survival, growth and aggressiveness but evidence as to the mechanisms of these effects is limited and often contradictory. Irrespective of nitrogen form, excessive use of nitrogenous fertilisers, especially at ill-judged stages in host growth, will result in increased losses from disease. This has been demonstrated for *Xanthomonas stewartii* on sweet corn; *Fusarium oxysporum* on *Chrysanthemum*, tomato and musk melon; *Verticillium dahliae* (section 2.6.9) on tomato; *V. albo-atrum* on hop; *Botrytis cinerea* (section 2.6.4) on *Chrysanthemum*, grapevine and strawberry; *B. gladiolorum* on *Gladiolus* and *B. allii* on onion. Applied at the correct growth stage, nitrogen may enable a host to survive pathogen attack. Dressings of nitrogen at 400 kg ha^{-1} are applied to overwintered cauliflower in spring to produce rapid foliage growth, thereby replacing leaves destroyed during the winter by infections of *Mycosphaerella brassicicola* (ringspot).

Potassium fertilisers are thought to reduce the effects of most pathogens particularly wilt-inducing fungi such as *Fusarium oxysporum* on tomato and cabbage (Walker and Foster, 1946). Potassium, however, appears to be without effect on *Verticillium* spp. and bacterial wilts of tomato (Gallegly and Walker, 1949; Walker *et al.*, 1954). Little detectable effect was found when potassium was applied to plants suffering from foot and root rots but, when associated with nitrogen fertiliser, potassium reduced the effect of *Fusarium culmorum* on wheat (Onuorah, 1969). Some reactions are more complex. With *Gibberella zeae* (maize stalk rot) increased potassium reduced disease when nitrogen level was low, had no effect at medium-nitrogen level and disease increased at high-nitrogen levels; with low levels of potassium, nitrogen was without effect (Thayer and Williams, 1960). High incidence of storage rot in carrot caused by *Mycocentrospora acerina* has been correlated with high levels of potassium in the field. The effects of root-rotting pathogens are generally reduced by high-phosphate nutrition. In Canada the reappearance of browning root rot of cereals caused by *Pythium* spp. has been attributed to insufficient use of phosphatic fertilisers (Vanterpool, 1962). Applications of phosphate have been shown to reduce damage caused by *Aphanomyces cochliodes* in beet.

The relationship of calcium nutrition to disease severity is even more tantalising since there are complex side-effects such as those on pH which may in turn lead to an imbalance of other nutrients especially trace elements. Studies have related soil calcium status to sodium levels. Increased sodium led to greater susceptibility of tomato to wilt (*Fusarium oxysporum* f. sp. *lycopersici*). This may be connected with a reduced availability of calcium in the presence of excess sodium. It has been clearly demonstrated that high calcium and hence alkaline pH values are related to reduced disease severity in tomatoes infected by *Verticillium albo-atrum* (wilt—section 2.6.9) (Hubbeling and Basu Chaudhary, 1969). Additional magnesium has been related to reduced infection of tomato by *Botrytis* and *Phytophthora* spp. and resulted in 50–60 per cent reductions in infection of beet by *Cercospora beticola* and *Pleospora betae* and beans by *Pseudomonas phaseolicola* (halo blight—section 4.5).

Soil and compost amendments may affect disease severity. Hardwood bark used as an amendment to peat composts controlled *Thanatephorus cucumeris* and *Pythium ultimum* the causes of crown and root rot of poinsettia (*Euphorbia pulcherrima*). Soil factors such as texture, porosity, water content and pedalogical composition influence the availability of nutrients and this effect in relation to disease development may be equally as important as the absolute quantity of each nutrient present in the soil. Although nutrients may be present in sufficient amounts, physical soil factors may lead to nutrient imbalances (excesses and deficiencies) which expose the crop host to pathogen invasion and colonisation. There may also be effects on plant vigour that in turn affect microclimate, thereby also influencing pathogen growth and sporulation. There are instances where rapid growth at the seedling stage may enable plants to escape infection. Conversely, rapid growth may be associated with etiolation, the formation of thin-walled parenchyma cells and subsequent invasion by downy mildew pathogens (Crute and Dixon, 1981).

Various systems of soil-less culture have been devised for the production of protected crops. These have the advantage of circumventing infection by root-invading pathogens which accumulate particularly where there is repeated use of border soil in protected structures. Initially use was made of compost beds from rotting farmyard manure and household refuse. For commercial horticulture this was unsuitable because of the limited quantities available, variable quality and high labour costs. An easier system was the use of straw bales which were placed in the glasshouse, saturated with water and, as decomposition began, covered with a thin layer of sterilised compost into which the crop plants were placed. The major difficulty with this method was to obtain straw guaranteed to be free from hormonal herbicide residues. From this sytem has developed the use of peat bags and bolsters in which crops like tomatoes and cucumbers are grown. Nutrients and water are applied continually during the season. Careful husbandry and regular chemical analyses are essential to maintain the nutrient-salt-moisture balances for crops which may grow for many months in an environment where the roots are restricted to a small volume of growing medium. Nonetheless these methods insulate the crop from a liability to infection by root-invading pathogens from the glasshouse soil, but there have been instances where infection has come from the peat. Complete freedom from soil and other solid substrates is obtained by the nutrient-film technique. Crops are grown in polyethylene or concrete troughs through which a steady trickle of nutrient solution passes, forming a film of moisture around the roots but permitting aeration. In the hands of a skilled grower this method can produce high yields but there must be constant monitoring and correction of pH and pC since all the buffering action of a solid rooting medium is absent. Nutrient-film technique is gaining acceptance in the UK whereas on the Continent greater use has been made of rockwool cubes. The plants are raised on these cubes and once germinated are transferred to polyethylene channels with each plant irrigated separately by tubes leading from a central nutrient source. Although such techniques circumvent the likelihood of invasion by traditional pathogens, new ones have emerged such as the upsurge in incidence of lettuce big vein (section 3.5.3) due to the easy spread of zoospores of the vector fungus *Olpidium brassicae* within the nutrient film.

9.1.2 Land use

In the field a primary cultural practice to control soil-borne pathogens is the use of rotations (Baker and Cook, 1974). Breaks of several seasons when a non-host is present in the soil dilutes soil-borne inoculum and permits the production of a susceptible crop for one or two seasons. The usefulness of this technique rests, however, on the longevity of soil-borne inoculum, width of pathogen host range and the ability of the pathogen to exist on plant residues. Direct evidence for a positive correlation between rotation and pathogen control is highly tenuous and arguments in favour of rotation or continuous monocropping for some field crops are finely balanced (Shipton, 1977). With horticultural crops produced on limited areas, however, the balance is in favour of rotation; for example, the incidence of *Plasmodiophora brassicae* (clubroot—section 2.6.1) was highest when cabbage was grown on the same land for three years. Production of brassicas after beet, carrot and grass crops decreased infection by up to 50 per cent with corresponding yield increases. Additionally it is now being realised that some at least of the 'replant disorders' that previously have prevented repeated growing of crops such as apple, pear, raspberry, rose and nursery stock are incited by pathogenic organisms. Sewell (1981) has implicated *Pythium sylvaticum* in the replant disorder of apples. The replant problem is acute with some nursery-stock crops. For instance, in order to produce *Thuja plicata* from seed regularly it is essential to rotate around 8–10 sites in order to avoid infection by *Keithia thujina* (needle blight). Rotations for vegetable crops are essential for the control of *Thanatephorus cucumeris* (damping-off) on brassicas or *Peronospora viciae* (downy mildew) and *Erysiphe pisi* (powdery mildew) on peas since these pathogens can perennate on crop residues some of which degrade only slowly in soil. Selection of suitable break crops will greatly influence the success of rotation for pathogen control since it is essential not to provide another host suitable to the pathogen. Control of *Leptosphaeria maculans* (brassica root rot and canker) is best achieved with lucerne and clover, *Aphanomyces cochlioides* (root rot) on beet is partially controlled by rotations that include maize, small-grain cereals and soy bean. Rotational techniques in warm temperate and semi-tropical areas are the only control measures available for bacterial pathogens such as *Pseudomonas solanacearum* and *Corynebacterium michiganense* on field-produced tomatoes.

Land-utilisation planning is an integral part of the proper use of rotational control since this can be affected by the distances that pathogen spores may be transmitted by rain or wind. Thus it is essential to keep seed beds well away from existing brassica crops to avoid infection by *Mycosphaerella brassicicola* (ringspot). Late summer field lettuce crops should not be situated near winter crops that have been established under glass to minimise the risk of infection by *Bremia lactucae* (downy mildew—section 2.6.3). Ware and seed crops should be well separated, especially in the case of biennial crops such as beet, to control *Peronospora farinosa* (downy mildew). For some pathogens the close proximity of alternative hosts must be avoided so as to discourage pathogen growth. Thus apples and raspberries should be well apart to minimise the risk of infection by *Leptosphaeria coniothyrium* (cane blight).

Under protected structures there is only limited scope for crop rotation particularly as there is a strong trend towards specialist production by monoculture of crops such as tomatoes, cucumbers or lettuces in order to obtain the highest financial returns. When labour costs were low, soil-borne pathogens could be controlled by literally removing them by annually changing the topsoil in the glasshouse. Soil removal was replaced by soil sterilisation with the development of steam-generating boilers. This form of sterilisation can be accomplished by introducing steam through perforated pipes placed up to 50 cm deep in the soil, pulling a steam plough consisting of a series of perforated pipes through the top 20–30 cm of cultivated soil, surface sterilisation by introducing steam beneath polyethylene sheeting laid on the soil, or the use of steam and air mixtures.

Costs of these techniques are now prohibitive in terms of fuel and labour requirements except for very specialised situations and steam sterilisation is replaced by chemical sterilisation (section 6.4.3). For some pathogens soil flooding which asphyxiates the propagules can be used as a control measure. This method is applied in Florida, USA, to control *Sclerotinia sclerotiorum*, a common and severe pathogen of many vegetable crops (Moore, 1949).

Crop disease may be avoided by advancing or delaying sowing or planting so that the crop and pathogen are out of phase with each other and no epidemic can develop. A traditional method for reducing powdery mildew (*Erysiphe cruciferarum*) infection on brassicas is to sow late; consequently the foliage is insufficiently mature to support this pathogen during the main summer period when the crop is at risk (Searle, 1920). A heavy price may be paid, however, in terms of reduced yield (Doling and Willey, 1969). Cabbage and other cole crops that are grown in colder seasons tend to suffer less from *Plasmodiophora brassicae* (clubroot—section 2.6.1) because the pathogen is less active, whereas summer cabbage and cauliflower crops are at high risk from this pathogen. Depth of sowing in relation to rate of emergence may be used to circumvent some diseases such as damping-off which is caused by a range of pathogens.

9.1.3 Crop habit

Crop growth habit may affect susceptibility to some pathogens; for example green beans (*Phaseolus vulgaris*) which produce pods held high off the ground are less prone to infection by *Colletotrichum lindemuthianum* (anthracnose) and *Botrytis cinerea* (grey mould—section 2.6.4). Similarly, standard roses are less affected by *Diplocarpon rosae* (black spot) than bush types (Gäumann, 1950). Pruning presents a means whereby pathogen invasion may be discouraged by reducing humidity in trees and bushes and foci of infection can be removed. Cutting out infected and cankered apple wood reduces the levels of *Podosphaera leucotricha* (powdery mildew), *Nectria galligena* (canker) and *N. cinnabarina* (coral spot) infection. Removal of mummified fruits during pruning reduces the inoculum potential for infection in the following season by pathogens such as *Sclerotinia fructigena* (apple brown rot and spur canker). *Didymella applanata* (spur blight of raspberry) is associated with overcrowding of new cane growth which should be thinned to 11 canes m^{-1} as soon after the end of picking as is practicable. Removal of individual infected buds may be practical for valuable ornamental specimens such as *Rhododendron* invaded by *Pycnostysanus azaleae* (bud blast). Pruning by clipping avoids blossom-time invasion of hawthorn (*Crataegus* spp.) hedges by *Erwinia amylovora* (fire-blight).

The entry of heart-rotting pathogens at pruning can be avoided by making clean pruning cuts and painting with wound-sealing compounds. Wounds caused by defoliating and side-shooting tomatoes need to be free from snags to reduce entry by *Botrytis cinerea* (grey mould—section 2.6.4). Buckeye rot caused by *Phytophthora nicotianae* var. *parasitica* on field-grown tomatoes in India was reduced by staking the plants and removing foliage and fruit up to 15-30 cm from soil level to prevent splash-entry of the pathogen (Sharma *et al.*, 1977). Twig die back caused by *Phomopsis pseudotsugae* on Japanese larch (*Larix leptolepis*) can be minimised by pruning with a saw rather than a bill hook, since a saw produces cleaner cuts and less bruising.

Ceratocystis fagacearum (oak wilt) spreads within stands from tree to tree; this can be prevented by poisoning trees around infected individuals or by undercutting between them to prevent root grafts. Early felling and deep girdling of infected individuals, especially in summer, may be used to control this pathogen. Tree stumps should either be removed or chipped and the chippings removed to prevent colonisation by butt-rotting fungi. Where this is impractical, immediate spraying with fungicides is essential. Trenching around trees or stumps may be used to inhibit spread of *Paxillus giganteus*, the fairy-ring fungus of pine trees, or *Armillaria* (*Armillariella*) *mellea* (honey fungus—section 2.6.7).

Frame-working of top fruit is advocated in place of top-grafting because there is less exposure of wood to infection by pathogens such as *Chondrostereum purpureum* (silver leaf). Additionally, frame-working brings the tree back into bearing more quickly. Bridge-grafting may be employed to save valuable trees badly infected by *Nectria galligena* (canker). Mixed planting may in some cases help to control pathogens; alternating rows of willow and poplar will reduce twig blight of willow caused by *Physalospora miyabeana.* Increasing the distance between plants can control some pathogens by decreasing their ease of spread. Losses due to *Sclerotinia sclerotiorum* (white mould) on *Phaseolus vulgaris* bean in Mexico were reduced by increasing plant-spacing. Increasing row-spacing from 50 to 95 cm raised yield from 796 to 1133 kg ha^{-1}.

9.1.4 Crop sanitation

Levels of pathogen inoculum can be greatly reduced by general sanitation such as removal of crop debris and eradication of ground-keepers after a crop has finished. Under glass these measures are particularly important with crops such as cucumber, lettuce, tomato, *Chrysanthemum* and carnation. In the field, removal, for example, of bulbs infected with *Botrytis tulipae* (tulip fire) limits secondary spread of the pathogen. Husbandry operations like chopping haulm to encourage rapid decomposition, and ploughing to 20–25 cm reduce spread of *Ustilago maydis* (sweet corn smut). Removal of debris from orchards and plantations where the crop has a life of several years is especially important in order to prevent build-up of infection and inoculum. *Pseudomonas syringae*, the cause of bacterial canker of cherry, is found on refuse and weeds in sour-cherry orchards in Michigan, USA.

Mechanical harvesting techniques have increased the possibilities of pathogen damage to crops and care must be exercised in setting up such machines correctly to minimise bruising and abrasion. For example, mechanical harvesting of carrots caused 20–30 per cent more damage to the roots than hand harvesting, and led eventually to proportionately greater rotting in store. Mechanical harvesters for raspberry damage the new season's growth while removing fruit from the old canes and this leads to greater invasion by *Leptosphaeria coniothyrium* (cane blight).

9.1.5 Biological control

Forms of biological control are beginning to be used against pathogens in some horticultural crops and this trend is likely to gather momentum in coming decades. Implantation of wooden dowels impregnated with *Trichoderma viride* into plum trees or application of a spore suspension to pruning wounds controls *Chondrostereum purpureum* (silverleaf) (Corke, 1980) and spray inoculation of pine tree stumps directly after felling with oidea of *Peniophora gigantea* gives good protection against *Heterobasidion annosum* (heart and butt rot) (Risbeth, 1963). Potentially *Verticillium lecani* may be used to control *Uromyces dianthi* (carnation rust) while *Pseudomonas* spp. inhibit the growth of *Puccinia allii* (leek rust) (Doherty and Preece, 1978) and *Penicillium brevicompactum, Trichoderma viride* and *Acremonium* spp. are strongly inhibitory to *Fulvia fulva* (tomato leaf mould) (Kashyap and Levkina, 1977). Extracts from higher plants may also be developed as means of biological control. Considerable activity is found in extracts from *Urginea ultissima* against *Phytophthora capsici.*

Some soils in the USA and France have been reported to suppress the growth of usually aggressive pathogens. In the USA a clay loam at Fort Collins, Colorado, possessed suppressive characteristics that inhibited infection of successive radish crops by *Thanatephorus cucumeris* (damping off). No correlation could be detected between suppressiveness of this soil and antagonism of the soil microflora as assayed *in vitro*. Suppressive soils tended, however, to possess greater lytic properties and populations of *Trichoderma* spp. than soils conducive to disease development. In France certain soils, particularly the alluvial soils of Chateau-Renard, are suppressive to the development of *Fusarium oxysporum* f. sp. *melonis* (melon wilt). Even when such soils are inoculated with *F. oxysporum* f. sp. *melonis,* no infection of susceptible melon plants takes place. The inhibitory effect persisted under a range of climatic and cultural conditions. It was transmitted to non-inhibitory soils that had been heat treated. Fungal inoculum placed in the soil persisted but showed no pathogenicity *in situ*. The inhibitory properties of soil persisted after heat treatment with steam: air mixtures at 50°C for 3 minutes but disappeared if the temperature was raised to 55°C (Rouxel *et al.*, 1977).

Cross protection

This occurs when avirulent strains of a pathogen prevent the development of virulent strains and is sometimes effective as a means of virus control (section 9.2.2) but there are also reports indicating that it may be of use against fungal pathogens. For example, incubated droplets containing spores of an avirulent race of *Colletotrichum lindemuthianum* (bean anthracnose) produced diffusates that prevented growth of virulent races (Berard *et al.*, 1972). The protection factor is effective on several host cultivars provided that they possess resistance genes similar to those of the donor. A cultivar carrying genes for resistance to two races of *C. lindemuthianum* produced two cross-protection factors that could be distinguished by their effects on cultivars bearing these genes separately (Berard *et al.*, 1973). This effect may eventually be used on a more widespread basis since inoculation of *Fusarium* spp. protected tomatoes against wilt caused by *F. oxysporum* f. sp. *lycopersici* (Homma and Ohata, 1977). Further, *Colletotrichum lindemuthianum* (bean anthracnose), from *Phaseolus* bean placed on cucumber inhibited disease caused by *Cladosporium cucumerinum* (scab) and *Colletotrichum lagenarium* (anthracnose) (Hammerschmidt *et al.*, 1976). There are also reports that inoculation of cucumber cotyledons with *C. lagenarium* provided protection against subsequent infection by the same pathogen.

Phytoalexins

Although so far largely studied for theoretical reasons, phytoalexins may eventually provide tools for the biological control of pathogens. These substances, which inhibit the growth of certain micro-organisms, are formed in higher plants in response to chemical, physical and biological stimuli. They can be detected at low concentrations in healthy non-infected plants and in larger quantities in some host–parasite combinations. Phytoalexins were first investigated during studies of the hypersensitive responses to *Phytophthora infestans* (potato blight—section 2.6.2) and later in experiments concerned with the reaction of bean pods to droplets containing spores of *Sclerotinia fructigena* (apple brown rot and spur canker) (Müller, 1959). A wide range of host families has since been shown to produce phytoalexins. Examples are: Compositae, Convolvulaceae, Leguminosae, Malvaceae, Orchidaceae, Solanaceae and Umbelliferae. An example of the effects of phytoalexins is seen in the work of Mansfield and Widdowson (1973) and Mansfield and Deverall (1974a, b) with *Botrytis* spp. and broad bean. *Botrytis fabae* (chocolate spot) is capable of colonising broad bean leaves, forming spreading lesions, whereas *B. cinerea* (grey mould—section 2.6.4) forms lesions only slowly, if at all. It is suggested that the resistance of broad bean leaves to species of *Botrytis*, other than *B. fabae*, is expressed following epidermal penetration and is associated with the formation of only limited lesions. Resistance operates by the restriction of lesion size within which growth of the invading hyphae is confined by about the third day after inoculation. The chemical that has been identified as most actively associated with lesion limitation in leaf and pod tissue is wyerone acid while wyerone accumulation is thought to inhibit growth of both *B. fabae* and *B. cinerea* in bean cotyledons. Results indicate that *B. fabae* can suppress phytoalexin production during the critical early phases of invasion and pathogen establishment. Factors that are important in the establishment or otherwise of a pathogen are: the number of cells killed by the fungus before phytoalexins accumulate, the sensitivity of hyphae to specific phytoalexins and the ability of hyphae to detoxify phytoalexins.

Although no practical means for pathogen control have yet resulted from these studies, knowledge of the ways in which invasion and colonisation are limited by host reaction may indicate directions in which crop protection could proceed (Deverall, 1977).

9.1.6 Integrated control

It is evident that no single pathway of crop protection will be totally successful over an extended period of time: cultivar resistance 'breaks down', pathogens become tolerant to agrochemicals and cultural methods are environmentally dependent. Consequently a mixture of techniques is required, which has been termed 'integrated control'. This concept may be defined as a pathogen management system which in the context of associated environmental and host and pathogen dynamics utilises all suitable techniques and methods in as compatible a manner as possible and maintains the pathogen population levels below those likely to cause economic injury. Use of such concepts requires not simply the juxtaposition or superimposition of all suitable management techniques—resistance, chemical, cultural, biological and plant health regulation—but also their integration. This philosophy for control demands far greater knowledge and understanding of host and pathogen for successful application on the part of those using control than has previously been the case where control was limited to use of a resistant cultivar or application of a fungicide. Strategies need to be developed not only for a particular host–parasite complex but for several and for their application and modification to specific holdings and seasons (Anon, 1968).

9.2 VIRUSES

No chemicals are available to control virus pathogens directly and host resistance has been exploited so far in only a few cases, for example, lettuce mosaic virus (section 3.5.2) (Ryder, 1970) and tomato mosaic virus (Pelham, 1966). Most techniques for control of viruses aim to control the vector rather than the pathogen and tend to be of a husbandry or biological form. The effects of viruses on crop yield are well established; for instance, paracrinkle virus in potato cv. King Edward reduced tuber weight by 10 per cent and caused tubers to be of more variable size. Removal of viruses from *Narcissus tazetta* cv. Grand Soleil d'Or produced more vigorous bulbs, forming greater numbers of larger flowers with more vibrant colour (Stone *et al.*, 1978); similarly virus-free *Lilium*, carnation, *Iris* and *Pelargonium* have shown improvements in flower quality when compared with infected material.

9.2.1 Virus-tested planting material

Virus-tested planting material may be obtained from either vegetatively propagated material or seed, although many viruses are not seed transmissible (Bennett, 1969; Zitter, 1977). Where a crop is produced from seed, substantial freedom from viruses that are seed transmissible is usually obtained by producing it in regions where vectors are virtually absent and then testing the seeds to ensure that only a minimal number is infected. The value of such seed has been demonstrated with lettuce mosaic virus (section 3.5.2). Lettuce crops grown from commercial seed batches containing from 2.2–5.3 per cent infection produced crops in which 25–96 per cent of plants were infected, whereas crops grown from virus-tested seed containing less than 0.1 per cent infected seed produced crops with 0.5 per cent or less infected lettuce plants (Broadbent *et al.*, 1951; Zink *et al.*, 1956; Purcifull and Zitter, 1971).

For vegetatively produced crops the simplest means of minimising virus infection is by roguing-out infected plants but this is only satisfactory where there are distinct and visually obvious symptoms and infected plants can be removed before much virus-spread has occurred. A more satisfactory system is to produce stocks of virus-tested planting material as with apple, pear, raspberry, strawberry, carnation, *Chrysanthemum* and *Narcissus*. An initial stock of virus-free material may be obtained by selection and indexing for virus-free individuals in commercial crops, heat therapy, meristem tip culture, or, more rarely, by cold treatment or chemotherapy. Once a virus-free stock is developed then a basic nucleus must be maintained in strict isolation to retain the pathogen-free state and to provide batches of material that may be increased to commercial quantities for sale as certified virus-tested material (figure 9.1 and section 8.2.1).

9.2.2 Prevention of spread into and within crops

Viruses can be spread over large distances by insect vectors and may also perennate in the absence of a host crop in wild plants. Weed hosts such as *Capsella bursa pastoris* (shepherd's purse), *Senecio vulgaris* (groundsel), *Stellaria media* (chickweed) and *Chenopodium album* (fat hen) are, for example, sources of beet mild yellowing virus which affects beet, spinach, lettuce and other crops (Russell, 1965). Some nematode-borne viruses, for example, tomato black ring virus persist for only a few weeks in the nematode vector, *Longidorus elongatus*, but for much longer in infected weed seeds. Hence when dormant weed seeds germinate in spring, virus-free *L. elongatus* may reacquire this virus from the roots of infected weed seedlings (Murant, 1970). Some crop plants may pose a danger to other crops; pea leaf roll virus is transmitted between lucerne, bean and pea crops by the aphid, *Acyrthosiphon pisum* (Smith, 1982). In some cases viruses cannot be transmitted within a crop but can only come from sources outside; for example, parsnip yellow fleck virus can be transmitted

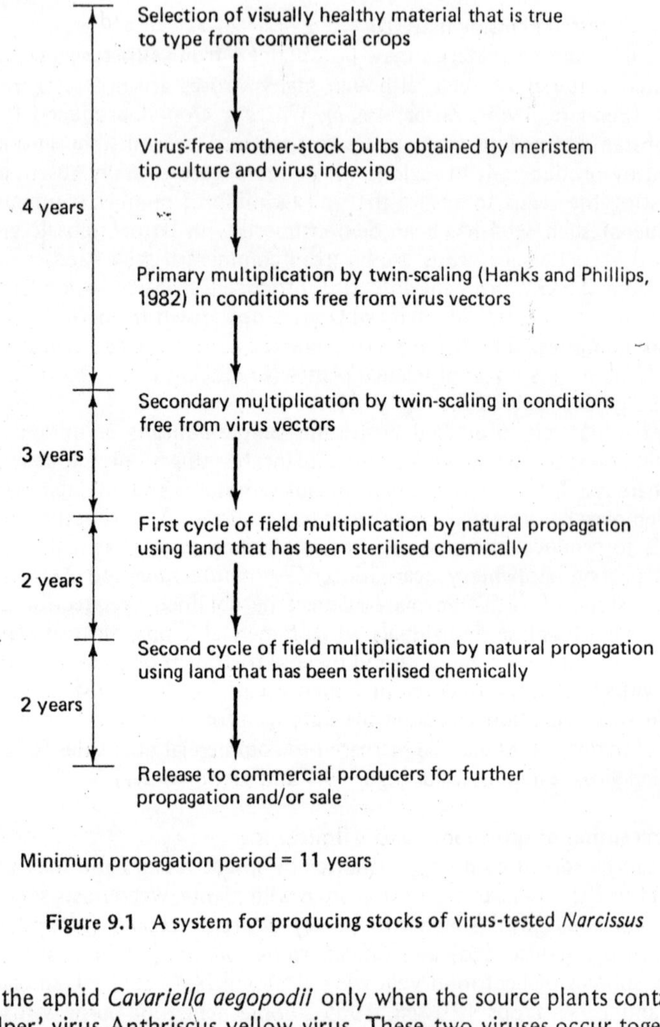

Figure 9.1 A system for producing stocks of virus-tested *Narcissus*

by the aphid *Cavariella aegopodii* only when the source plants contain the 'helper' virus Anthriscus yellow virus. These two viruses occur together in wild umbellifers but not in parsnip; hence transmission of parsnip yellow fleck virus is from wild umbellifers to parsnip but not from parsnip to parsnip (Murant, 1974). The ingress of virus from outside a crop may also take place from storage clamps containing crops such as cabbage, carrot and potato, or from biennial seed crops which act as reservoirs of virus for newly sown crops in spring.

Control of virus infection may be achieved by techniques such as crop rotation and cultivation, isolation by distance, isolation by time, interplanting barrier and cover crops, and by the effects of field size and crop density. For example, since most aphids land at the edges of fields, larger fields tend to diminish the incidence of infection. Crop density has a similar effect; experiments have shown that 51 per cent of sugar beet plants were infected with beet mild yellows virus with a crop density of 17 500 plants ha^{-1} whereas only 15 per cent were infected with a crop density of 126 500 plants ha^{-1} (Heathcote, 1974). Some plants are more susceptible to virus infection when juvenile, becoming more resistant with age. Consequently, sowing or planting dates may be advanced or retarded to ensure that the host is sufficiently mature to resist infection when vectors are most active. This technique has been used in California, USA, to control celery viruses (Stevens and Nienow, 1947; Milbraith, 1948). In New Zealand, enforcement of a crop-free period has eliminated onion yellow dwarf virus (Bos, 1976). Towards the end of some crops the aerial parts may be mechanically or chemically destroyed to prevent virus infection; this is practised regularly in the Netherlands to protect potato seed crops from infection where there are only limited opportunities for geographical isolation. Alternatively, seed or vegetative stock crops may be grown in cool, windy or possibly wet isolated areas where the insect vectors arrive late in the season. Barriers such as the use of strips of barley around cauliflower seed beds may be employed to inhibit infestation by vectors. In Israel straw mulches are used to repel *Trialeurodes vaporariorum* (whitefly) vectors of bottle gourd mosaic virus in cucumber (Nitzany *et al.*, 1964). Aluminium strips may be laid in crops to reflect ultraviolet light, thereby repelling aphid vectors. This latter method when used in *Gladiolus* crops produced a 95 per cent reduction in aphid numbers present and cucumber mosaic virus damage fell by two-thirds. Aluminium foil has been used in Florida, USA to reduce the spread of water melon mosaic virus in squash crops (Moore *et al.*, 1965). Similarly, plastic mulches can provide protection. Following wing development, aphids are strongly attracted by the shorter wavelengths of light and fly towards the sky; after several hours this effect is reversed and they are repelled by short wavelengths and attracted to yellow and green, therefore searching for plant material (Johnson, 1969). Consequently sheets of yellow polyethylene laid outside the field will attract winged aphids and have been used in California, USA to reduce the spread of cucumber mosaic virus, potato virus Y and tomato yellow leafcurl virus. Yellow sticky boards are used successfully in glasshouses to control *Trialeurodes vaporariorum* (whitefly) (Smith *et al.*, 1979).

Control of viruses has been achieved by inoculation of plants with avirulent strains of virus. Deliberate inoculation of tomato seedlings with the avirulent mutant M-II-16 has been used as a cultural method to minimise losses in yield and fruit quality caused by tomato mosaic virus (Rast, 1972). There are disadvantages from this technique since inoculation with mild virus strains may itself reduce yield by 5–10 per cent. The infected crop can act as a reservoir of virus from which other species or cultivars may become infected. Mild or avirulent strains of virus may mutate to more severe strains in some hosts, and serious disease may result from mixed infections when unrelated viruses are introduced into a crop. Use has also been made of grafting techniques whereby the virus resistance present in wild species of *Cucurbita* and *Lycopersicon* is utilised to protect otherwise susceptible but high-yielding commercial cultivars (Maxon-Smith and Proctor, 1965). The resistant wild species is used as a rootstock and the commercial cultivar is used as a scion. Seedlings of both are raised in the normal way and when the plants are of sufficient size to handle the scion is bench-grafted on to the rootstock. When the graft union is complete the scion cultivar can be severed from its own root system and grown throughout the cropping period on the roots of the resistant stock. For cucumbers *C. ficifolia* is used as a rootstock while tomato rootstocks have been developed as F_1 hybrids between *L. esculentum* x *L. hirsutum* which have the added bonus of increasing yields possibly because of the marked heterosis associated with the F_1 hybrid rootstocks.

Pesticides may be used to protect crops from virus pathogens by killing or inhibiting the vectors. Requirements for such chemicals must include action that is rapid enough to prevent virus acquisition and transmission, and with minimal development of resistance to the pesticide by vector populations. Other requirements are similar to those outlined in section 6.1 for crop-protection chemicals. Materials that are used include nicotine, derris, pyrethrum, the synthetic pyrethroids, organochlorine and organophosphorus insecticides. Efficient use of these materials requires a trapping organisation to monitor the build-up of vector populations, such as the suction traps operated throughout the UK by Rothamsted Experimental Station with the aid of other organisations, integration of this information with meteorological data and the issue of spray warnings to growers in order to obtain maximum efficacy. To combat soil-borne fungal-transmitted and nematode-transmitted viruses, sterilant chemicals such as carbon disulphide, formaldehyde, methyl bromide, chloropicrin, DD mixture, Telone (DD plus additional dichloropropene) and PCNB may be used (section 6.4.3). PCNB is nematicidal, as well as fungicidal, and will inhibit *Longidorus elongatus* which can infect strawberry with tomato black ring virus and raspberry ringspot virus (section 3.5.4). Problems associated with soil sterilisation to control soil-borne viruses are related to the fact that nematodes move slowly in soil but penetrate deeply. Trials have shown that DD or methyl bromide will kill more than 99 per cent of *Xiphinema diversicaudatum* and prevent the infection of strawberry with Arabis mosaic virus. With this level of nematode elimination, virus infection was reduced by 97 per cent but when 90 per cent of nematodes were killed infection was reduced by only 56 per cent. *Olpidium brassicae* the fungal vector of lettuce big vein (section 3.5.3), may be controlled by soil fumigation with DD and chloropicrin. The development of organo carbamate and organo phosphate systemically active pesticides in granular formulations has meant that they can be applied at, or immediately after, drilling and provide longer-lasting protection for the crop.

Man may mechanically transmit viruses such as tomato mosaic virus by movement in the tomato crop and on knives used for side-shooting and defoliation. Dipping knife blades in 3 per cent trisodium orthophosphate eliminates the latter mode of transmission, but spread may still occur on hands and clothes (Broadbent, 1976). Application of oil, or oil:water emulsions, to foliage interferes with virus transmission from the vector to the crop; this was first reported by Bradley *et al.* (1962) and has been reviewed by van der Veken (1977). Commercial use of this method is inhibited by inadequate efficacy and phytotoxicity. Recently, specially formulated mineral oil (JMS-Stylet Oil) has been used to control virus infection in squash and tomato without phytotoxicity (Zitter and Ozaki, 1978). Requirements for such oils and their use are a high degree of chemical inertness and viscosity, and precise application techniques. In Israel the incidence of cucumber mosaic virus in plots of cucumber sprayed with 5–10 per cent oil emulsions at low volume was reduced to 10–20 per cent of that in untreated plots. In the Netherlands weekly applications of mineral oils are used to protect virus-tested *Lilium* cultivars (Asjes, personal communication). The big advantage of this method is lack of mammalian toxicity. Other materials that inhibit transmission of virus from vector to host plant include plant lipids, milk fat and silicones.

Herbicides may be employed to remove weed hosts and groundkeepers which act as reservoirs of virus infection (Duffus, 1971). There are also some instances where predators and parasites of the vector may be used to prevent virus transmission. Carrot motley dwarf virus is not epidemic in California, USA, as it is in Australia because the aphid vector is heavily parasitised by Braconid wasps which are absent in Australia. Introduction of the predator *Aphidius* sp. to the Melbourne area of Australia eliminated the aphid vector of carrot motley dwarf virus and it ceased to cause crop losses in this region.

So far, despite intensive research, no commercial antiviral chemicals are available to inhibit either infection by virus particles or their multiplication. Quarantine techniques may be used to prevent the long-range spread of viruses from country to country (section 8.1).

REFERENCES

ANON (1968). Report of the first session of the Food and Agriculture Organisation (FAO) panel of experts on integrated pest control. *Food and Agriculture Meeting Report*, no. PL/1967/M/7, FAO, Rome

BAKER, K. F. and COOK, R. J. (1974). *Biological Control of Plant Pathogens*, Freeman, California

BENNETT, C. W. (1969). Seed transmission of plant viruses. *Advances Virus Research*, 14, 221–61

BERARD, D. F., KUĆ, J. and WILLIAMS, E. B. (1972). A cultivar-specific protection factor from incompatible interactions of green bean with *Colletotrichum lindemuthianum*. *Physiological Plant Pathology*, 2, 123–7

BERARD, D. F., KUĆ, J. and WILLIAMS, E. B. (1973). Relationship of genes for resistance to protection by diffusates from incompatible interactions of *Phaseolus vulgaris* with *Colletotrichum lindemuthianum*. *Physiological Plant Pathology*, 3, 51–6

BLAKEMAN, J. P. (1981). The chemical environment of the leaf surface in relation to growth of pathogenic fungi. 255–68, in *Ecology of Leaf Surface Microorganisms* (eds T. F. Preece and C. H. Dickinson), Academic Press, London

BLOOM, J. R. and COUCH, H. B. (1958). Influence of pH, nutrition and soil moisture on the development of large brown patch. *Phytopathology*, 48, 260 (abstract)

BOS, L. (1976). *Onion Yellow Dwarf Virus*. Commonwealth Mycological Institute/ Association of Applied Biologists. Descriptions of Plant Viruses No. 158. Commonwealth Mycological Institute, Kew

BRADLEY, R. H. E., WADE, C. V. and WOOD, F. A. (1962). Aphid transmission of potato virus Y inhibited by oils. *Virology*, 18, 327–8

BROADBENT, L. (1976). Epidemiology and control of tomato mosaic virus. *Annual Review Phytopathology*, 14, 75–96

BROADBENT, L., TINSLEY, T. W., BUDDIN, W. and ROBERTS, E. T. (1951). The spread of lettuce mosaic in the field. *Annals Applied Biology*, 38, 689–706

COLHOUN, J. (1953). Observations on the incidence of clubroot disease of *Brassicae* in limed soils in relation to temperature. *Annals Applied Biology*, 40, 639–44

COLHOUN, J. (1961). Spore load, light intensity and plant nutrition as factors influencing the incidence of clubroot of *Brassicae*. *Transactions British Mycological Society*, 44, 593–600

COOPER, A. J. (1956). The influence of cultural conditions on the development of *Alternaria* leaf spot of cinerarias. *Journal Horticultural Science*, 31, 229–33

CORKE, A. T. K. (1980). Biological control of plant pathogens. *Report Long Ashton Research Station for 1978*, 139–41

COUCH, H. B. and BLOOM, J. R. (1960). Influence of environment on diseases of turf grasses. II. Effect of nutrition, pH and soil moisture on *Sclerotinia* dollar spot. *Phytopathology*, 50, 761–3

CRUTE, I. R. and DIXON, G. R. (1981). Downy Mildew Diseases caused by the Genus *Bremia* Regel. 420–60, in *The Downy Mildews* (ed. D. M. Spencer), Academic Press, London

DALY, J. M. (1949). The influence of nitrogen source on the development of stem rust of wheat. *Phytopathology*, 39, 386–94

DAS, A. C. and WESTERN, J. H. (1959). The effect of inorganic manures, moisture and inoculum on the incidence of root disease caused by *Rhizoctonia solani* Kuhn in cultivated soil. *Annals Applied Biology*, 47, 37–48

DEVERALL, B. J. (1977). *Defence Mechanisms of Plants*. Cambridge Monographs in Experimental Biology No. 19, Cambridge University Press, London

DOHERTY, M. A. and PREECE, T. F. (1978). *Bacillus cereus* prevents germination of uredospores of *Puccinia allii* and the development of rust disease of leek, *Allium porrum*, in controlled environments. *Physiological Plant Pathology*, 12, 123–32

DOLING, D. A. and WILLEY, L. A. (1969). Date of sowing and the yield of swedes. *Experimental Husbandry*, 18, 87–90

DUFFUS, J. E. (1971). Role of weeds in the incidence of virus diseases. *Annual Review Phytopathology*, 9, 319–40

ENDO, R. M. (1966). Control of dollar spot of turf grass by nitrogen and its probable basis. *Phytopathology*, 56, 877 (abstract)

GALLEGLY, M. E. and WALKER, J. C. (1949). Plant nutrition in relation to disease development. V. Bacterial wilt of tomato. *American Journal Botany*, 36, 613–23

GÄUMANN, G. (1950). *Principles of Plant Infection* (translation of *Pflanzliche Infektionslehre* by W. B. Brierley), Crosby Lockwood and Son, London, p. 259

HAMMERSCHMIDT, R., ACRES, S. and KUĆ, J. (1976). Protection of cucumber against *Colletotrichum lagenarium* and *Cladosporium cucumerinum*. *Phytopathology*, 66, 790–3

HANKS, G. and PHILLIPS, S. (1982). Twin-scaling. *Growers Bulletin No. 6*, Glasshouse Crops Research Institute, Littlehampton

HEATHCOTE, G. D. (1974). The effect of plant spacing, nitrogen fertiliser and irrigation on the appearance of symptoms and spread of virus yellows in sugar beet crops. *Journal Agricultural Science, Cambridge*, 82, 53–60

HOMMA, Y. and OHATA, K. (1977). Suppression of *Fusarium* wilt symptoms in tomato by prior inoculation of other formae of *F. oxysporum* and *F. solani*. *Agricultural Experiment Station, Shikoku*, Bulletin No. 30, 103–14

HUBBELING, N. and BASU CHAUDHARY, K. C. (1969). Influence of pH and calcium nutrition on the resistance of tomato seedlings to *Verticillium* wilt. *Mededelingen Rijksfakulteit Landbouwweetenschappen, Gent*, 34, 937–43

HUBER, D. and WATSON, R. D. (1974). Nitrogen form and plant disease. *Annual Review Phytopathology*, 12, 139–65

JOHNSON, C. G. (1969). *Migration and Dispersal of Insects by Flight*, Methuen, London

KASHYAP, V. and LEVKINA, I. (1977). Effect of micro-organisms isolated from tomato leaves on mycelial growth of some pathogenic fungi. *Vestnik Moskovskogo Universiteta Biologiya*, No. 1, 65–9

KATAN, J. (1981). Solar heating (solarization) of soil for control of soil borne pests. *Annual Review Phytopathology*, 19, 211–36

LAST, F. T. (1953). Some effects of temperature and nitrogen supply on wheat powdery mildew. *Annals Applied Biology*, 40, 312–22

LI, C. Y., LU, K. C., TRAPPE, J. M. and BOLLEN, W. B. (1967). Selective nitrogen assimilation by *Poria weirii*. *Nature, London*, 213, 814

MANSFIELD, J. W. and DEVERALL, B. J. (1974a). The rates of fungal development and lesion formation in leaves of *Vicia faba* during infection by *Botrytis cinerea* and *Botrytis fabae*. *Annals Applied Biology*, 76, 77–89

MANSFIELD, J. W. and DEVERALL, B. J. (1974b). Changes in wyerone acid concentrations in leaves of *Vicia faba* after infection by *Botrytis cinerea* or *B. fabae*. *Annals Applied Biology*, 77, 227–35

MANSFIELD, J. W. and WIDDOWSON, D. A. (1973). The metabolism of wyerone acid (a phytoalexin from *Vicia faba* L.) by *Botrytis fabae* and *B. cinerea*. *Physiological Plant Pathology*, 3, 393–404

MAUDE, R. B. (1970). The control of *Septoria* on celery seed. *Annals Applied Biology*, 65, 249–54

MAXON-SMITH, J. W. and PROCTOR, P. (1965). Use of disease resistant rootstocks for tomato crops. *Experimental Horticulture*, 12, 6–20

MILBRAITH, D. G. (1948). Control of western celery mosaic. *California Department Agriculture Bulletin*, 37, 3–7

MOORE, W. D. (1949). Flooding as a means of destroying sclerotia of *Sclerotinia sclerotiorum*. *Phytopathology*, 39, 920–7

MOORE, W. D., SMITH, F. F., JOHNSON, G. V. and WOLFENBARGER, D. O. (1965). Reduction of aphid populations and delayed incidence of virus infection on yellow straight neck squash by use of aluminium foil. *Proceedings Florida State Horticultural Society*, 78, 187–91

MÜLLER, K. O. (1959). Hypersensitivity. 469–519, in *Plant Pathology–An Advanced Treatise*, Vol. I, *The Diseased Plant* (eds. J. G. Horsfall and A. E. Dimond), Academic Press, New York

MURANT, A. F. (1970). *Tomato Black Ring Virus*. Commonwealth Mycological Institute/Association of Applied Biologists. Descriptions of Plant Viruses No. 38. Commonwealth Mycological Institute, Kew

MURANT, A. F. (1974). *Parsnip Yellow Fleck Virus*. Commonwealth Mycological Institute/Association of Applied Biologists. Descriptions of Plant Viruses No. 129. Commonwealth Mycological Institute, Kew

NILSSON, G. I. and NELSON, P. V. (1964). Nitrogen nutrition and development of *Phialophora cinerescens* in carnation. *Phytopathology*, 54, 1172–3

NITZANY, F. E., GEISENBERG, H. and KOCH, B. (1964). Tests for the protection of cucumbers from a white fly-borne virus. *Phytopathology*, 54, 1059–61

ONUORAH, P. E. (1969). Effect of mineral nutrition on the *Fusarium* brown foot-rot of wheat. *Plant Soil*, 30, 99–104

PELHAM, J. (1966). Resistance in tomato to tobacco mosaic virus. *Euphytica*, 15, 258–67

PURCIFULL, D. E. and ZITTER, T. A. (1971). Virus diseases affecting lettuce and endive in Florida. *Proceedings Florida State Horticultural Society*, 84, 165–8

RAST, A. Th. B. (1972). M II-16, an artificial symptomless mutant of tobacco mosaic virus for seedling inoculation of tomato crops. *Netherlands Journal Plant Pathology*, 78, 110–12

RISBETH, J. (1963). Stump protection against *Fomes annosus*. III. Inoculation with *Peniophora gigantea*. *Annals Applied Biology*, 52, 63–77

ROTEM, J. and COHEN, Y. (1966). The relationship between mode of irrigation and severity of tomato foliage diseases in Israel. *Plant Disease Reporter*, 50, 635–9

ROUXEL, F., ALABOUVETTE, C. and LOUVET, J. (1977). Research on the resistance of soils to disease. II. Effect of heat treatment on the microbiological resistance of a soil to vascular *Fusarium* disease of melon. *Annals Phytopathologie*, 9, 183–92

RUSSELL, G. E. (1965). The host range of some English isolates of beet yellowing viruses. *Annals Applied Biology*, 55, 245–52

RYDER, E. J. (1970). Inheritance of resistance to common lettuce mosaic. *Journal American Society Horticultural Science*, 95, 378–9

SEARLE, G. D. (1920). Some observations on *Erysiphe polygoni* DC. *Transactions British Mycological Society*, 6, 274–94

SEWELL, G. W. F. (1981). Effects of *Pythium* species on the growth of apple and their possible causal role in apple replant disease. *Annals Applied Biology*, 97, 31–42

SHARMA, S. L., CHOWFLA, S. C. and SOLI, H. S. (1977). Control of buckeye rot of tomato by cultural practices. *Indian Journal Mycology Plant Pathology*, 6, 51–4

SHEPHARD, M. C. and WOOD, R. K. S. (1963). The effect of environment and nutrition of pathogen and host in the damping off of seedlings by *Rhizoctonia solani*. *Annals Applied Biology*, 51, 389–402

SHIPTON, P. J. (1977). Monoculture and soilborne plant pathogens. *Annual Review Phytopathology*, 15, 387–407

SMITH, F. F., WEBB, R. E. and DUDLEY, R. F. (1979). Greenhouse whitefly control by sticky yellow board traps. Abstract 654, *9th International Congress Plant Protection*, Washington, DC

SMITH, K. M. (1972). *A Textbook of Plant Virus Diseases*, Longman, London

SOL, H. H. (1967). The influence of different nitrogen sources on (1) the sugars and amino acids leached from leaves and (2) the susceptibility of *Vicia faba* to attack by *Botrytis cinerea. Mededelingen Landbouwhogeschool Opzoekingsstations*, Gent, 32, 768-75

STEVENS, N. E. and NIENOW, I. (1947). Plant disease control by unusual methods. *Botanical Review*, 13, 116-24

STONE, O. M., BRUNT, A. A. and HOLLINGS, M. (1978). Methods, logistics and problems in the production, distribution and use of virus-free clones of *Narcissus tazetta* cv. Grand Soleil d'Or. *Report Glasshouse Crops Research Institute for 1977*, 149-67

STRIJBOSCH, Th. (1976). Relationship between the occurrence of condensation and botrytis. *Annual Report Glasshouse Crops Research and Experiment Station Naaldwijk*, 1973 and 1974, 79-80

THAYER, P. L and WILLIAMS, L. E. (1960). Effects of nitrogen, phosphorus and potassium concentrations on the development of *Gibberella* stalk and root-rot of corn. *Phytopathology*, 50, 212-14

VAN DER VEKEN, J. J. (1977). Oils and other inhibitors of non-persistent virus transmission. 435-54, in *Aphids as Virus Vectors* (eds K. F. Harris and K. Maramorosch), Academic Press, New York

VANTERPOOL, T. C. (1962). *Pythium* root rot of wheat in Saskatchewan. *Canadian Plant Disease Survey*, 42, 214-15

WALKER, J. C. (1946). Soil management and plant nutrition in relation to disease development. *Soil Science*, 61, 47-54

WALKER, J. C. and FOSTER, R. E. (1946). Plant nutrition in relation to disease development. III. *Fusarium* wilt of tomato. *American Journal Botany*, 33, 259-64

WALKER, J. C. and WELLMAN, F. L. (1926). Relation of temperature to spore germination and growth of *Urocystis cepulae. Journal Agricultural Research*, 32, 133-46

WALKER, J. C., GALLEGLY, M. E., BLOOM, J. R. and SHEPHERD, R. D. (1954). Relation of plant nutrition to disease development. VIII. *Verticillium* wilt of tomato. *American Journal Botany*, 41, 760-2

WEINHOLD, A. R., DODMAN, R. L. and BOWMAN, T. (1972). Influence of exogenous nutrition on virulence of *Rhizoctonia solani. Phytopathology*, 62, 278-81

ZINK, F. W., GROGAN, R. G. and WELCH, J. E. (1956). The effect of the percentage of seed transmission upon subsequent spread of lettuce mosaic virus. *Phytopathology*, 46, 662-4

ZITTER, T. A. (1977). Epidemiology of aphid borne viruses. 385-412, in *Aphids As Virus Vectors* (eds K. F. Harris and K. Maramorosch), Academic Press, New York

ZITTER, T. A. and OZAKI, H. Y. (1978). Aphid-borne vegetable viruses controlled with oil sprays. *Proceedings Florida State Horticultural Society*, 91, 287-9

FURTHER READING

BAILEY, J. A. and MANSFIELD, J. W. (1982). *Phytoalexins*, Blackie, Glasgow

COLHOUN, J. (1973). Effects of environmental factors on plant disease. *Annual Review Phytopathology*, 11, 343-64

COSTA, A. S. (1976). Whitefly-transmitted plant diseases. *Annual Review Phytopathology*, 14, 429-49

HARDISON, J. R. (1976). Fire and flame for plant disease control. *Annual Review Phytopathology*, 14, 355–79

PALTI, J. (1981). *Cultural Practices and Infectious Crop Diseases*, Springer-Verlag, Berlin

PALTI, J. and ROTEM, J. (1981). Control of Downy Mildews by Cultural Practices. 289–304, in *The Downy Mildews* (ed. D. M. Spencer), Academic Press, London

STEVENS, R. B. (1960). Cultural practices in disease control. 357–429, in *Plant Pathology—An Advanced Treatise*. Vol. III. *The Diseased Population, Epidemics and Control* (eds J. G. Horsfall and A. E. Dimond), Academic Press, New York

SUMNER, D. R., DOUPNIK, B. and BOOSALIS, M. G. (1981). Effects of reduced tillage and multiple cropping on plant diseases. *Annual Review Phytopathology*, 19, 167–87

ZENTMYER, G. A. and BALD, J. G. (1977). Management of the environment. 121–44, in *Plant Disease—An Advanced Treatise*, Vol. I. *How Disease is Managed* (eds J. G. Horsfall and E. B. Cowling), Academic Press, New York

ZITTER, T. A. and SIMONS, J. N. (1980). Management of viruses by alteration of vector efficiency and by cultural practices. *Annual Review Phytopathology*, 18, 289–310

10 CONTROL: POST-HARVEST PATHOGENS

Storage of sound high-quantity horticultural produce curtails its exposure to air and soil-borne pathogens in the outdoor environment and should therefore reduce losses. Pathogens present as latent infections on produce going into storage and more frequently new infections from air-borne pathogens present within a store will cause disease on crops as they are stored. Disease development is encouraged by biochemical and physiological changes which begin once crops are harvested. A primary aim of storage should be to limit such changes for as long as possible. *Clamps* and *barn* types of storage provide cool conditions for preserving commodities such as cabbage, carrots, celery, onions and potatoes. The economic margins to be gained by storage are often such as to permit the use of more sophisticated techniques, rather than these traditional methods, whereby losses due to pathogens may be reduced. Storage may be *short-term*—a few days—as for soft fruit and loose leafy vegetables in order to remove *field heat*, thereby delaying the onset of wilting and disease and permitting transit over long distances; this is achieved by *vacuum, forced air, hydro or ice-bank cooling*. Storage may be *long-term*—up to several months—by means of *refrigeration* or *controlled atmospheres*.

10.1 PATHOGENS ASSOCIATED WITH POST-HARVEST LOSSES

There are three main types of post-harvest deterioration: physical, physiological and biochemical, and pathological. Fungi causing post-harvest decay are usually species of *Alternaria, Botrytis, Diplodia, Monilinia, Penicillium, Phoma, Phomopsis, Rhizopus* or *Sclerotinia*; bacteria causing similar decay are species of *Erwinia* and *Pseudomonas*. Viruses may also be responsible for losses in store; for instance, turnip mosaic virus is associated with the development of large necrotic lesions in stored Dutch white Langedijk cabbage. Numbers of pathogenic types responsible for post-harvest losses are few compared with those infecting standing crops. Mostly they are weak parasites that largely invade previously injured tissue or tissue affected by metabolic changes. Some species of *Alternaria, Botrytis, Monilinia* and *Penicillium* will, however, penetrate undamaged tissue. Injury due to hand or machinery damage is frequently also a major site for invasion (Apeland, 1974; Tucker, 1974), and so is cracking that has resulted from wide fluctuations in temperature or water supply while the crop has been in the field (Davies, 1977).

10.2 OBJECTIVES OF STORAGE

Storage is usually achieved by some form of *refrigeration* or *natural cooling* which retards the following.

1. *Respiration and other metabolic activity.*
2. *Ageing as a result of ripening, softening, textural and colour changes.*
3. *Moisture loss and resultant wilting.*
4. *Direct spoilage as a result of growth of post-harvest pathogens.*
5. *Undesirable growths such as potato sprouts.*

Growth and invasion by decay-inducing micro-organisms are either completely stopped or slowed significantly at 0°C. Since the respiration rate and rate of deterioration of produce are interrelated, treatments that reduce respiration also slow down deterioration. Techniques such as hydro-cooling may be used for commodities such as asparagus, celery, radish, sweet corn and topped carrots. For some produce there may be a requirement for *pre-cooling* prior to hydro-cooling to prevent tissue damage, which will act as sites for pathogen entry, by sudden exposure to water near to freezing-point and it is necessary to include *sterilant chemicals* in the water to prevent an accumulation of micro-organisms. Vacuum-cooling removes moisture from produce, thereby leading to loss of turgor and possible pathogen entry. Control of relative humidity is essential for successful storage of horticultural produce; the RH should be at least 90-95 per cent thereby retarding shrivelling but not encouraging pathogen growth (van den Berg and Lentz, 1966; Apeland and Baugerød, 1971). Apples and pears will withstand refrigeration to 0°C whereas vegetables such as aubergines, bell peppers, cucumbers, green beans, squash and tomato are injured by temperatures below 10°C, leading to accelerated pathogen-induced deterioration (Eckert and Sommer, 1967).

10.3 SUPPLEMENTARY TECHNIQUES

10.3.1 Careful handling and protective packaging

Cuts and bruises increase water loss and admit pathogens and also detract from quality *per se*. The use of tray packs, liners, plastic bags and covers reduces such damage. Other important factors are proper loading patterns both in store and in transit to permit rapid air circulation and the removal of noxious gases such as ethylene.

10.3.2 Inducing wound barriers

For produce such as potatoes, pre-conditioning at 15.5–21 °C permits the formation of the peridermal layer and suberisation of cells, thereby inhibiting fungal invasion (Smith and Smart, 1956).

10.3.3 Modified atmospheres

In controlled atmosphere stores it is usual to decrease oxygen and increase carbon dioxide concentrations; this reduces respiration rate, thereby delaying ripening and physiological deterioration, and retarding growth by fungi and bacteria. Such storage is used world-wide for apples. Initially it overcame the difficulties associated with storage of apple cultivars which were susceptible to cold damage in refrigeration and prevented physiological disorders such as soft scald, brown core and soggy breakdown, which could all lead to pathogen-induced rotting. Atmospheres are changed to contain 2–3 per cent oxygen and 1–5 per cent carbon dioxide, depending on the cultivar involved and storage system. High concentrations of carbon dioxide (20–30 per cent) are now used for cherries and strawberries for short-term storage because levels of carbon dioxide above 10 per cent will retard the growth of most pathogens (Brown, 1922). Often this is accomplished by placing solid carbon dioxide ('dry ice') in transit boxes (Harvey, 1967). Controlled-atmosphere storage is occasionally employed for vegetables; in the USA it is utilised for long-distance transport of lettuce using 2 per cent carbon dioxide atmospheres. Use is also being made of atmospheres containing high levels of nitrogen to prolong produce life and to prevent decay.

10.3.4 Hypobaric storage

Storage at reduced pressures such as 0.25 kg cm^{-2} retards ripening of cherries and green tomatoes and inhibits decay in strawberry (Tolle, 1972).

10.3.5 Manipulation of relative humidity

Irrespective of temperature, relative humidity at 90 per cent and above is essential to prevent produce desiccation. To this end shrivelling of apple, pear and carrot may be prevented by using plastic liners within bulk storage bins (Hardenburg, 1974). Decay due to raised relative humidity does not increase until RH rises above 95 per cent. It is largely the condensation of moisture on produce surfaces, not high RH, that enhances decay. Loss of turgor renders carrots susceptible to damage by grey mould caused by *Botrytis cinerea* (section 2.6.4); while the roots are turgid the secondary phloem parenchyma is capable of localising fungal invasion. An 8 per cent loss of fresh weight in carrots predisposed them to infection by *Rhizopus stolonifer* and susceptibility to *Botrytis cinerea* increased with 5–10 per cent losses in fresh weight.

10.3.6 Heat treatment

This was used initially to control decay in green tomatoes (Rosenbaum, 1920). More recently, treatment with hot humid air has been used in Hawaii and Australia, and hot water has been used in the USA and elsewhere. Dips in water at 54°C for 2.5 minutes destroy surface pathogens and those beneath the produce-skin such as *Monilinia* and *Rhizopus* on peach and nectarine (Smith, 1962). Exposure of melons to hot water at 54°C for 0.5 minutes was especially effective in controlling stem end rots, while treatment with water-saturated air at 40°C for 1 h controlled decay in figs, raspberry and strawberry (Smith and Worthington, 1965; Worthington and Smith, 1965; Couey and Follstad, 1966). In France this technique has been applied to fresh market produce but it can lead to storage injury if treated produce is subsequently placed in controlled-atmosphere stores. Hot-water treatments are faster and cause less injury than hot air because the temperature rise is localised to the produce surface and there is no necessity for long post-treatment cooling periods; the process is similar to domestic bottling techniques. Addition of fungicides to the hot water permits a reduction in temperature and the amount of chemical required is less than that needed at normal temperatures to control pathogens.

Onion bulbs are heat treated in Hungary by placing them in the sun for 12 days to inhibit decay caused by *Peronospora destructor*; disease is prevented if the temperature reaches 41°C for 4 h. Artificial heat-drying of onions in the UK reduces neck rot induced by *Botrytis allii* by curing the bulbs at 32–34°C in store. This dries out the onion neck tissues which are the main avenue for infection especially if the onion leaves are removed in the field by flailing. Similarly, curing *Gladiolus* corms at 35°C for 14 days reduces the liability to infection by *Botrytis gladiolorum.*

10.4 APPLICATION OF CHEMICALS AFTER HARVEST

The use of post-harvest chemicals should be viewed as ancillary to successful mechanical storage. Usually they are required because of some inadequacy in the storage system or because of the ability of a few important pathogens to grow at temperatures below those at which it is convenient to store particular fruit or vegetables. The main criterion for post-harvest crop-protection chemicals (in addition to those listed in section 6.1 for chemicals generally) is long residual activity although with reduced pathogen metabolism during storage there is increased time for a material to operate. Phytotoxicity should not be a problem provided that there is no scalding action by the material. Some cases of tolerance by pathogens to post-harvest chemicals have been reported in the USA (Harding, 1959); for example various *Penicillium* spp. tolerant to biphenyl and sodium orthophenyl phenate and tolerance to benomyl, thiabendazole and 2-aminobutane in *P. digitatum* and *P. italicum*. In artificial culture, tolerance to 2,6-dichloro-4-nitroaniline has been identified in *Rhizopus stolonifer*, *Botrytis cinerea* and *Gilbertella persicaria*. Chemicals cleared for use in the UK on stored vegetables are 2-amino butane, benomyl, dichloran, dichlorophen, non oxymol-iodine complex and thiabendazole. One of the crops most frequently treated after harvest is Dutch white cabbage which is stored extensively in Europe for marketing until June following harvest in October–November. This product will store for 3–4 months at ambient winter temperatures in northern Europe and 8–10 months under refrigeration. Weight losses are 1.0–1.5 per cent per month, owing to shrivelling plus losses due to pathogen damage and physiological breakdown. The crop must be stored before it is overmature or frosted and handled carefully during harvesting and storage. Rots develop progressively in store and it may be necessary to trim off the wrapper leaves several times (Derbyshire and Shipway, 1978). Dusting the heads with quintozene has been used in the Netherlands but is now questioned on health grounds. Drenches with benomyl and thiabendazole will control grey mould caused by *Botrytis cinerea* (section 2.6.4) but will increase damage caused by *Alternaria* spp. and *Pseudomonas marginalis*. Similarly, dust treatment with thiabendazole and dichloran in bins at 1-3°C for 7 months controlled *B. cinerea* but had no effect on *Alternaria brassicicola*. *Phytophthora porri* (the cause of white tip of leek) has been identified as a cause of progressive head rot of cabbage which starts at the cut stem end (Geeson, 1976). Infections develop from soil-splash or contaminated knives particularly during wet harvests. Internal pepper-spotting of this crop may be reduced by controlled atmosphere storage at 5 per cent carbon dioxide and 1 per cent oxygen. Frequently the application of chemicals pre-harvest is a more efficient means of controlling post-harvest pathogens such as *Botrytis*, *Gloeosporium*, *Penicillium* and *Rhizopus* spp. on fruit (Powelson, 1960; Burchill and Edney, 1963). *Botrytis* blight on cherry is not a significant pathogen prior to harvest In California, USA, but blossom-stage sprays will prevent incipient infections from developing on the green fruit. Sprays during blossoming are essential for grapes to prevent post-harvest losses due to grey mould caused by *B. cinerea* (Ogawa *et al*., 1976).

General sanitation

Strict cleanliness in handling produce for storage and transit will reduce losses caused by decay and careful removal of rotten produce avoids spread by removing foci of infection. Empty stores should be cleaned by introduction of ozone which removes storage odours (Schomer and McColloch, 1948; Spalding, 1966). If allowed to accumulate these might subsequently increase the rate of ripening of produce. Ozone also kills pathogen spores. Cleaning boxes with calcium hypochlorite and packing lines with steam will reduce future losses.

REFERENCES

APELAND, J. (1974). Storage quality of carrots after different methods of harvesting. *Acta Horticulturae*, 38, 353-7

APELAND, J. and BAUGERØD, H. (1971). Factors affecting weight loss in carrots. *Acta Horticulturae*, 20, 92-7

BROWN, W. (1922). On the germination and growth of fungi at various temperatures and in various concentrations of oxygen and carbon dioxide. *Annals Botany*, 36, 257-83

BURCHILL, R. T. and EDNEY, K. L. (1963). The control of *Gloeosporium album* rot of stored apples by orchard sprays which reduce sporulation of wood infections. *Annals Applied Biology*, 51, 379-87

COUEY, H. M. and FOLLSTAD, M. N. (1966). Heat pasteurization for control of post harvest decay in fresh strawberries. *Phytopathology*, 56, 1345-7

DAVIES, W. P. (1977). Infection of carrot roots in cool storage by *Centrospora acerina*. *Annals Applied Biology*, 85, 163-4 (abstract)

DERBYSHIRE, D. M. and SHIPWAY, M. R. (1978). Control of post-harvest deterioration in vegetables in the UK. *Outlook on Agriculture*, 9, 246-52

ECKERT, J. W. and SOMMER, N. F. (1967). Control of diseases of fruits and vegetables by post harvest treatment. *Annual Review Phytopathology*, 5, 391-432

GEESON, J. D. (1976). Storage rot of white cabbage caused by *Phytophthora porri*. *Plant Pathology*, 25, 115-16

HARDENBURG, R. E. (1974). Use of plastic films in maintaining quality of fresh fruits and vegetables during storage and marketing. *Symposium Semi-Annual Meetings American Society of Heating, Refrigerating and Air-Conditioning Engineers, Chicago, ASHRAE Bulletin*, CH-73-7, 19-29

HARDING, P. R. (1959). Biphenyl-induced variations in citrus blue mould. *Plant Disease Reporter*, 43, 649-53

HARVEY, J. M. (1967). Modified atmospheres in transport. *United Fresh Fruit and Vegetable Association Yearbook*, pp. 193-6

OGAWA, J. M., MANJI, B. T. and EL-BEHADLI, A. H. (1976). Chemical control of post harvest diseases. 561-75, in *Proceedings 3rd International Biodegradation Symposium* (eds J. M. Sharpley and A. M. Kaplan), Applied Science Publishers, London

POWELSON, R. L. (1960). Initiation of strawberry fruit rot caused by *Botrytis cinerea*. *Phytopathology*, 50, 491-4

ROSENBAUM, J. (1920). Infection experiments on tomatoes with *Phytophthora terrestrea* Sherb. and a hot water treatment of the fruit. *Phytopathology*, 10, 101-5

SCHOMER, H. A. and McCOLLOCH, L. P. (1948). Ozone in relation to storage of apples. *United States Department of Agriculture Circular* No. 765, 1-24

SMITH, W. L. (1962). Reduction of post harvest brown rot and rhizopus decay of eastern peaches with hot water. *Plant Disease Reporter*, 46, 861-5

SMITH, W. L. and SMART, H. F. (1956). Relation of soft rot development to protective barriers in Irish potato slices. *Phytopathology*, 45, 649-54

SMITH, W. L., and WORTHINGTON, J. T. (1965). Reduction of post harvest decay of strawberries with chemical and heat treatments. *Plant Disease Reporter*, 49, 619-23

SPALDING, D. H. (1966). Effect of ozone on appearance and decay of strawberries, peaches and lettuce. *Phytopathology*, 56, 586 (abstract)

TOLLE, W. E. (1972). Hypobaric storage of fresh produce. *United Fresh Fruit and Vegetable Association Yearbook*, pp. 27–43

TUCKER, W. G. (1974). The effect of mechanical harvesting on carrot quality and storage performance. *Acta Horticulturae*, 38, 359–72

VAN DEN BERG, L. and LENTZ, C. P. (1966). Effects of temperature, relative humidity and atmosphere composition on changes in quality of carrots during storage. *Food Technology*, 20 (7), 104–7

WORTHINGTON, J. T. and SMITH, W. L. (1965). Post harvest decay control of red raspberries. *Plant Disease Reporter*, 49, 783–6

FURTHER READING

DALRYMPLE, D. G. (1967). The development of controlled atmosphere storage of fruit. *United States Department Agriculture Federal Extension Service Market Utilization Services*

EXKERT, J. W. (1967). Application and use of post-harvest fungicides. 287–378, in *Fungicides, An Advanced Treatise*, Vol. I, *Agricultural and Industrial Applications Environmental Interactions* (ed. D. C. Torgeson), Academic Press, New York

ECKERT, J. W. (1978). Post-harvest diseases of citrus fruits. *Outlook on Agriculture*, 9, 225–32

ISENBERG, F. M. R. (ed.) (1977). Symposium on vegetable storage. *Acta Horticulturae*, no. 62

LUTZ, J. M. and HARDENBURG, R. E. (1968). The commercial storage of fruits, vegetables, and florist and nursery stocks. *United States Department of Agriculture* Handbook No. 66, 1–43

LYONS, J. M. (1973). Chilling injury in plants. *Annual Review Plant Physiology*, 24, 445–66

PHILLIPS, W. R. and ARMSTRONG, J. G. (1967). The storage environment. *Handbook on the storage of fruits and vegetables—for farm and commercial use*. Canada Department of Agriculture Publication No. 1260

SMITH, W. L. (1976). Non-chemical control of post-harvest deterioration of fresh produce. 577–87, in *Proceedings of the 3rd International Biodegradation Symposium* (eds J. M. Sharpley and A. M. Kaplan), Applied Science Publishers, London

STOLL, K. (1974). The storage of vegetables in controlled atmospheres. *International Institute Refrigeration*, 54, 1302–24

TOMKINS, R. G. (1966). The choice of conditions for the storage of fruits and vegetables. *Annual Report East Malling Research Station for 1965*, 60–76

TUGWELL, B. L. (1973). Post-harvest fungicides. *Journal Australian Institute Agricultural Science*, 39, 167–73

HOST INDEX

MAIN AUTHORITIES

BAILEY, L. H. (1961). *Manual of Cultivated Plants*, Macmillan, New York.
CLAPHAM, A. R., TUTIN, T. G. and WARBURG, E. F. (1959). *Excursion Flora of the British Isles*, Cambridge University Press, London.

MICRO-ORGANISM AND DISEASE INDEX

(Page numbers in *Italic* type indicate detailed treatment)

GENERAL INDEX

(*indicates that this term is used repeatedly—the page indicated provides a definition)